D1525335

Wonder

Wonder

Childhood and the Lifelong Love of Science

Frank C. Keil

The MIT Press
Cambridge, Massachusetts
London, England

The MIT Press would like to thank the anonymous peer reviewers who provided comments on drafts of this book. The generous work of academic experts is essential for establishing the authority and quality of our publications. We acknowledge with gratitude the contributions of these otherwise uncredited readers.

This book was set in Stone Serif and Stone Sans by Westchester Publishing Services. Printed and bound in the United States of America.

Library of Congress Cataloging-in-Publication Data

Names: Keil, Frank C., 1952– author.
Title: Wonder : childhood and the lifelong love of science / Frank C. Keil.
Description: Cambridge, Massachusetts : The MIT Press, [2022] |
Includes bibliographical references and index.
Identifiers: LCCN 2021007025 | ISBN 9780262046497 (hardcover)
Subjects: LCSH: Cognition in children. | Reasoning in children. | Intuition. |
Science—Study and teaching.
Classification: LCC BF723.C5 K424 2022 | DDC 155.4/13—dc23
LC record available at https://lccn.loc.gov/2021007025

10 9 8 7 6 5 4 3 2 1

For Kristi

Contents

Preface

When our children were young, we lived in a house on the edge of a cliff overlooking Cayuga Lake, in Upstate New York. As we went up and down the ninety-eight (!) stairs to the lake, little bits of the shale cliff would break off and land on the stairs, much of the time revealing small fossils: shells, trilobites, leaves, and other mysterious patterns etched in stone. These attracted the curiosity of all our children making their way up and down the stairs as toddlers. They soon noticed that those patterns implied the presence of something more than an inanimate rock, something that suggested life. They would ask, "What is that?" "Why is it there?" "What happened to it?" and "How did it last so long?" Each of our attempts to provide an answer usually triggered a follow-up question. In those days before search engines, we'd run to our multivolume 1929 *Encyclopedia Britannica* we'd bought at a yard sale and look for answers, or sometimes we'd call relevant colleagues at Cornell. Our entire family fondly remembers those times even now more than thirty years later. As we will see, a tremendous amount of cognitive activity was occurring in those seemingly simple events, activity that reveals children's powerful propensities for learning about their world.

This book describes a convergence of major advances in recent research on children's minds. Those advances reveal an extraordinary cluster of early emerging abilities that help explain this youthful wonder and joy of discovery. We see preschoolers and even infants as driven to learn not just facts and images of their world but also underlying causal patterns that are at the very heart of science. Infants start with certain foundational contrasts between broad domains, such as the social and the nonsocial, that become greatly elaborated during the preschool and early school years. They learn not just as individuals but also as members of knowledge communities and show impressive early senses of how to best "harvest" knowledge from these communities and how to best engage with other minds.

Yet those joyous youthful moments of discovery and engagement often fade as children grow older, cease to wonder about the world, and stop building new understandings. I describe in detail how this decline occurs and argue that it is neither inevitable nor desirable. The new findings about children's natural cognitive inclinations show how an early love of science can be sustained throughout one's life with enormous attendant benefits. When we no longer enjoy wondering and searching for answers, we fail to expand our grasp of the world. We may never learn even the most basic causal processes that give insight into the workings of devices and the natural world. We digest food and unlock doors with no grasp of what happens within that makes digestion and unlocking possible. Even worse, we often are not even aware of how little we know and become ever more vulnerable to misunderstanding and manipulation by others.

In a world of ever more dizzying discovered and engineered complexity, it might seem that we all have to surrender our agency and just passively rely on others without any real sense of what they know and why. I argue against such a view in the three parts of this book.

First, I describe the extraordinary cognitive abilities of young children and preverbal infants. I show how these young humans have a diverse set of talents for making conjectures about how things work and how they are driven to expand on their initial understandings through engaging in individual explorations, through accessing what others know and by sensing what kinds of causal patterns and structures are likely to be especially informative. Young children possess many of the critical precursors to mature scientific thought as well as the passion for exploration and discovery that we associate with model scientists.

Second, I show how the glorious rise of wonder in the early years can be stifled and distorted into paths where the desire to discover fades away. I will describe seductive but mistaken views of young minds and their motivational processes. I then show how these misunderstandings can lead to disastrous declines of spontaneous exploration, conjecture, and questioning during the early school years and beyond. None of these are inevitable, but they happen all too often.

Finally, in the third part, I describe how suppressing wonder leads to drastically different futures from cases in which it is encouraged and supported.

I'll describe bleak lives of disengagement from science accompanied by distrust and denialism. I'll show how these orientations can be fueled by cognitive biases and motivated intellectual blind spots. Ultimately, such stances toward science lead to dogma and distortion that corrupt our understandings of all that is around us. I'll then describe a different path illuminated by the lives of a few inspirational individuals who never stop wondering and reveling in the successive discoveries that made their daily experiences ever deeper and more rewarding. I'll show how those positive cognitive habits that we all possessed as children can be reawakened to benefit not just ourselves but also society at large. By better understanding how illusions and social influences can derail us and make us more susceptible to cognitive biases and negative attitudes toward science, we can see how different, more rewarding paths are also possible. We can become more like children, continuously experiencing a genuine joy in discovery and preferring activities that enable us to better grasp causal structure to simply being attracted to the mere appearances of pretty things.

When we reawaken these early cognitive habits, we stop being alienated from science, ignorant of basic models, and sometimes openly hostile to the scientific community. By building on children's natural cognitive proclivities, we can avoid traps that lead to alienation and ignorance. We become much more able to combat the ever-rising tide of disinformation and to participate more fully in important policy decisions relating to science and technology. In the broadest sense, I argue that reengagement with science is critical to our individual and collective futures.

Acknowledgments

I am deeply grateful to many people for helping me improve this book from the earliest stages through several full-length drafts.

My most profound debt is to Kristi, my wife and soulmate for forty-nine years. Kristi has discussed almost every aspect of this book with me at length and has read several drafts of the full-length manuscript and many outlines. She has encouraged me on countless occasions when the project seemed overwhelming. She has persistently, but gently, reminded me of my tendency to write run-on sentences and convoluted passages, but somehow always pointing out these rambling disasters with humor and support. She also is my most important research collaborator having led several of the studies described in this book and been a key player in many others. Her brilliance, her experimental skills and her extraordinarily perceptive intuitions about children have helped me understand and see much more clearly how children's minds grow. Her very presence, with her incredibly lively and creative mind, constantly inspires me to new paths of wonder. Most of all, however, I am infinitely grateful for her ability to put up with me while being the most gifted parent on the planet. Through her, I learned more about children than from all other sources combined. Through her, I saw how a parent can encourage and nourish wonder. Through her, I learned desperately needed information about how to be a better parent myself. With her, I was able to witness the growth to adulthood of our three sons, Derek, Dylan, and Marty, the most rewarding experiences of my life by far. And most recently, I'm experiencing all this again as I watch her with our granddaughters Frances, Maren, and Poppy.

I am also deeply thankful to each of our three sons, not just as ideal exemplars of lifelong wonderers but because of their contributions to this project. Derek, the chemistry major and now architect, has talked with me on many occasions about many of the topics in this book, each time with incredible insight and suggestions for further lines of inquiry. His curiosity about the workings of the world remains fierce and unbounded. Dylan, the engineering major and start-up wizard, is an utterly insightful reader and commentator on just about everything, seeing patterns that almost everyone else misses. His comments on drafts of the prospectus and the full-length book were exactly on target and incredibly resonant with some of the leading experts in the field. Marty, the biology major, Teach for America high school science teacher, PhD in science education, and doctor of medicine, knows more about learning and engaging with science from more perspectives than anyone I know. His comments on drafts and his many suggestions arising from conversations about so many topics in this book have made all the difference. Our three sons have all given me deeply appreciated insights and support.

I am also delighted that Eden, Derek's wife and mother of Frances, and gifted artist and editor, has offered a great many perceptive observations about topics in this book. Kate, Dylan's wife, superstar geoscientist and recent mother of Maren and Poppy, somehow found the time to make incredibly helpful comments on each chapter of this book, always with an acute eye to celebrating science.

Beyond my family, my gratitude is broad and deep. I cannot thank enough my dear friend and colleague Paul Bloom. For years, Paul has been urging me to write a more accessible book along these lines and, through his comments on successive outlines, patiently guided me through the process of writing a book for a much broader audience. His comments on a full-length draft of this manuscript were profoundly helpful and insightful. He has been unrelentingly encouraging throughout, seeing in me much more potential than I really had, but helping me get at least closer to that ideal.

I have been inspired by and learned much from Dick Brodhead, a remarkable scholar, teacher, and academic leader at the national level. I wanted to know how to connect my ideas to those who don't specialize in the sciences, and I could not have found a better tutor than this leading figure in the humanities. Dick schooled me on the origins and meanings of wonder and helped me see ways to connect to much broader streams of thought. His comments on my book, and on versions of the title, were invaluable.

I am also ever in debt to another leading scholar in the humanities, Katie Lofton, a riveting scholar of religion and history and an inspirational dean of humanities at Yale. In a chance hallway encounter, I mentioned I was working on this book and she cheerfully asked to see it. I sent a late draft assuming that in the midst of the COVID-19 crisis it would vanish under piles of correspondence about remote teaching, social distancing and departmental affairs. Yet, somehow, only a few weeks later, Katie provided extraordinarily helpful comments that greatly influenced the final draft.

I thank Laura Kang, current principal of East Woods School, for spending many hours going through school archives so as to provide me with fascinating details about my remarkable science teacher more than half a century earlier. Many thanks as well to Vavara Petrolova for her knowledge about the Sputnik era from a Russian perspective and for researching Soviet archives that I could never access or read.

I will always be ever in debt to all past and present members of my extended lab group from postdocs to undergraduates to lab managers to summer interns. There are too many to list here, but they have all been enormously important colleagues who have shaped my ideas in so many different ways over the years. I simply could not have traveled this road of research and scholarship for the past four decades alone and acknowledge here their many contributions. Whenever I describe a study using the pronoun "we", I am indicating the involvement of others, often in key roles. Their names are well represented in the many endnotes.

I thank three anonymous reviewers of an early stage of this book, whose detailed suggestions on every chapter were invaluable in guiding my later writing. I am also especially indebted to two reviewers of the near final draft manuscript who made immensely helpful and detailed comments on every chapter.

Some of the material in this book is based on work supported by the National Science Foundation under (NSF grant DRL 1561143). Any opinions, findings, and conclusions or recommendations expressed in this material are those of the author and do not necessarily reflect the views of the National Science Foundation. Within NSF, I express special thanks to the EHR Core Research (ECR) program and its visionary and passionate leader Gregg Solomon. I am also grateful to the National Institutes of Health, the Templeton Foundation, and the McDonnell Foundation for funding my research and interdisciplinary meetings. As with NSF, my writings here do not necessarily reflect the views of those funders. I am especially grateful to two inspirational leaders of the McDonnell Foundation, John Bruer and Susan Fitzpatrick, who for more than thirty years have provided me with incredible opportunities to acquire important new perspectives on cognitive science, learning and education.

Finally, I thank many people at the MIT Press. Ever since my undergraduate days at MIT, the Press has been an essential part of my career growth as a cognitive scientist. Press director Amy Brand and senior editor Phil Laughlin have been longtime colleagues in other projects, and I was delighted when they expressed an interest in this book. Phil has been brilliant in his choice of the anonymous reviewers and in his feedback on various drafts. I am especially grateful to Deborah Cantor-Adams, associate managing editor, and to Regina Gregory for clarifying and insightful edits. I am also grateful to the many contributions of the other members of the magnificent MIT Press team: Alex Hoopes, assistant acquisitions editor; Molly Seamans, designer; Sean Reilly, art coordinator; Tori Bodozian, production coordinator, and Heather Goss, publicist. Finally, many thanks to my thoughtful indexer, Tobiah Waldron.

I The Cognitive Gifts of Childhood

1 The Puzzle and the Promise

The Rise and Fall of Wonder

A few miles from our house a celebrated independent bookstore carries a large collection of new books and old classics. What books do the store's well-educated clients buy? Roughly half the top sellers are nonfiction. On the nonfiction list, usually, at most, one book focuses on science. In national science best-seller lists, books in the top twenty rarely explain mechanisms, arguably the core product of science. Instead, we find biographies and self-help guides. Highly ranked "science" books often turn out to be attacks on science, branding research as deceitful, racist, misogynist, and corrupt. In October 2019 the ten bestselling "science and math" books on such lists included a diet regime, an analysis of how media falsely slandered the US president, and a guide on how to age more slowly. All of these last three books may have provided valuable information, but they definitely did not provide insights into underlying mechanisms or causal principles. The most traditional science topic covered in the top ten was a Cat in the Hat book on the solar system meant for kindergartners. Expand the list to the top twenty science and math books and we find an analysis of White people's difficulties talking about racism, techniques of psychotherapy, a guide for how to unleash our infinite potential and how to take a journey with the untethered soul. To be sure, other books in the top twenty did focus on more traditional science topics such as the history of humans, social and cognitive psychology, and a beautiful "illustrated exploration" of facets of chemistry. Even then, surprisingly little content provides causal explanations of the natural and engineered worlds. More broadly, in the top ten best-selling books in all areas for the past ten years on some lists, only one book out of 100 would be considered to be on a typical science topic: Neil deGrasse Tyson's *Astrophysics for People in a Hurry*. Even the title seems to suggest that we wouldn't want to linger on the topic.

In contrast, children's books for those under age five at the same bookstore often embrace science. One such book is the *National Geographic Little Kids First Big Book of Why*, full of questions such as "How do seeds grow?" "Why do balloons float?" and "How do glasses work?" More systematically, the top ten children's nonfiction books intended for children five and under listed in May 2020 included *What If You Had Animal Teeth?*; *Hidden Figures*; *Why Do Kittens Do That? Real Things Kids Love to Know*; *We Are the Gardeners*; *The Backyard Bug Book for Kids: Storybook, Insect Facts, and Activities*; and *Cool Cars and Trucks* (a book about how to build them). In a nearby toy store, popular toys for young children include magnifying boxes for inspecting insects, magnet sets, telescopes, and various construction kits. Almost anywhere one looks, children seem to have a hunger for scientific explanations and adults have little appetite if any.

Perhaps the age difference in orientation simply reflects parental aspirations for children's future careers, but the children's behaviors in both stores suggest otherwise. Well before they enter their first classroom, children eagerly explore their environment in search of answers to an endless stream of *why* and *how* questions.

A different indication of the age difference in preferences for content can be found in book titles. Search through the 160 million books in US Library of Congress for titles that start with certain phrases that are strongly linked to seeking causal explanations, such as "Why are . . ." and "How do . . ." In October 2019, roughly 250 books had titles starting with "Why are . . ." Most of these were adult books about nonscience topics, such as *Why Are Artists Poor? The Exceptional Economy of the Arts*, and *Why Are Contractors Always Late?* Roughly twenty-five books were clearly about topics in the natural sciences, and twenty-four of those were meant for children, with titles such as *Why Are Animals Different Colors?* and *Why Are There Waves?* The one adult exception was *Why Aren't Black Holes Black?*, a 1997 book on unanswered questions in science.

The adult "why are" books are often intriguing and insightful; they just aren't about traditional areas of science. The same pattern holds for other title beginnings. For example, roughly 1,100 books begin with "How do . . ." Of those, approximately one tenth are about engineering and science (e.g., *How Do Airplanes Fly?* and *How Do Animals Move?*). In that tenth, the vast majority (well over 90 percent) are children's books. The much larger number of books, not about science and engineering, are mostly adult focused (e.g., *How Do Banks Manage Liquidity Risk?* or *How Do Churches Grow*?).

This sampling of book titles in a massive library converges with the age-related differences in best-selling books. Children seem to be more interested in the hows and whys of the world than are adults. A burst of research in the last decade confirms the apparent pattern in book titles.[1] From three to six years of age, children ask many sincere how and why questions,[2] but then those questions plummet during the elementary school years and beyond.[3] Older children may ask questions more efficiently, but the sheer number of why and how questions drops dramatically.

Most children's spontaneous love of science fades by adulthood. Is it a problem with science itself? Does increasing familiarity with science and technology lead to disenchantment? Despite some claims to that effect, the real cause is a decline in wonder. Wonder is the engine that drives exploration and discovery, and, when it disappears, an infatuation with the workings of the world melts away.

We are all born with many essential ingredients of wonder—with inquisitive minds, fascinated by the world around us. Young children and even infants are naturally engaging in intuitive science every day, often with sophisticated methods. But that early bonfire of inquiry can shrink to a tiny flicker. This loss of wonder is not because we suddenly understand everything—we don't—but because distrust, disengagement, and denial can become embedded into many aspects of our lives. The consequences of this loss of wonder are profound. Because impoverished wonder can lead to especially poor understandings of underlying mechanisms, we become vulnerable to misinformation and manipulation by others. Even worse, abandonment of wonder deprives us of the intensely rewarding joy of discovery. This loss of wonder, however, is not inevitable. By better understanding its emergence and flowering in childhood and by recognizing the competing forces we face as adults, we can all take simple actions to reawaken that initial spark and live lives illuminated by wonder. The scientist, physician, and astronaut Mae Jemison, who showed a precocious and lifelong passion for science, captured the centrality of wonder beautifully in a 2019 interview:

> For me, I wanted to know how the world works. Every child wants to know how the world works. . . . That's the point at which we really need to connect . . . : honoring and respecting and integrating that incredible curiosity. Children come out picking up bugs and looking at things and trying to experiment, and that's how they learn about the world. . . . one of the issues that can happen in education is we actually destroy that [curiosity]. We want a specific answer. We ask for something to be memorized. And that's not the best way. . . . our goal is to harness that innate construct for information gathering and problem-solving and to refine it.[4]

In 1979, the psychologist Margaret Donaldson published a landmark book straightforwardly entitled *Children's Minds*.[5] She showed through clever experiments and elegant arguments how rigid stage theories were mistaken. Young children were not trapped in conceptually shallow early stages of thought that made some forms of understanding and insight unavailable. This book was one of the first and most powerful arguments leading to a revolution in the study of cognitive development. It is also an affirmation of an early drive to wonder. Near the end of the book, Donaldson described this early drive as plummeting when children enter school. In a later paper, she described the problem as arising from a "mismatch between school and children's minds."[6] Unfortunately, the mismatch and its corrosive effects on wonder continue, just as Donaldson implied. As I write this just a few months after Donaldson's death at age ninety-four, I am hoping we may finally confront the mismatch.

Wonder

The word *wonder* first appeared in Old English around AD 675 in a hymn by Cædmon, an allegedly illiterate cow herder from what is now southern Scotland.[7] Cædmon's wonder (wundor or uundra in Old English) implied an awe of the natural world, including humans. Awe remains a theme in many uses of wonder up to the present, but wonder in our sense means something more than awe. Awe alone often is passive—verging on the meaning of dumbstruck or stupefied reverence. We'll use the more active meaning of wonder described by the *Oxford English Dictionary* (OED) as "desirous to know or to learn." Wonder is a drive that comes from within. Richard

Holmes, known for his biographies of romantic poets and his book *The Age of Wonder: How the Romantic Generation Discovered the Beauty and Terror of Science*, describes wonder as having varied meanings that keep evolving over the ages, but he constantly returns to the idea of a childlike passion to know and explore. Changes in in the meaning of wonder often include shifts in moral implications. For example, historians of science Lorraine Daston and Katherine Park describe a dizzying array of dramatic changes in meaning and moral tone from 1150 to 1750.[8]

Today, wonder has different senses concerning its manner of engagement with science. Lisa Sideris, a scholar of religion, science, and the environment, describes one view of wonder as the means for revealing with stark clarity the story of the universe. Sideris sees this view as leading to arrogance and a feeling of human mastery over the world.[9] She favors instead a sense embodied in the naturalist's Rachel Carson's writings, one of humility and increasing mystery as each new discovery revealed even deeper puzzles.[10] Given such multiple senses and connotations, we need to specify the particular meaning intended here, a meaning that is closely connected to how young children learn about the world.

Wonder is not the same as curiosity.[11] Curiosity often means a desire to know "What's next?" or "What is that missing thing?" Curiosity about facts is also different from the sense of wonder intended here. Curiosity about facts often leads to a simple answer and then a dead end. For example: Q. What state has the most cars? A. California. (That ends the discussion.) Curiosity about cause and mechanism has cognitive legs and is closer to what we mean here by wonder. Most why and how questions lead to a potentially endless stream of follow-ups. Q. How do sunflowers move? A. Sunflowers turn their faces during their growth phase. They turn toward the sun in the morning and track it through the day. Q. How does that work? Q. Why does that occur only in growing sunflowers? A. . . . Causal mechanistic questions open up new territories of explanation. They arouse a sense of embarking on exciting journeys of discovery. Facts can sometimes do this as well by implying some kind of expanding information, but frequently they do not. Here, "wonder" describes a quest for understanding, which usually involves asking how and why. This is the most central motivating factor behind a lifelong interest in science. Instead of merely asking "What's next?" wonder also asks "Is it something like X or something like Y, or Z that is next?"

Wondering as intended here goes beyond merely being curious. Building on prior knowledge and some sense of major causal and spatial patterns, we entertain rough sketches of possibilities or interpretations and strive to learn which is more accurate and how it is filled out. To marvel at something is also linked to wondering. While marvels and wonders often refer to awe-evoking things, when children engage in wonder, they do much more than simply sit in a state of passive reverential awe.[12] Their awe is better described by Carson—a joyous marveling at how an insight has revealed an enormous new expanse of possible patterns to explore. It is not the dumbstruck, potentially fear-laden sense of awe experienced by adults. Almost a century ago, in an isolated region of Papua-New Guinea, the anthropologist Margaret Mead observed that, when children were asked to explain why a canoe tied to a tree drifted away overnight, they offered explanations of how the rocking boat gradually loosened up the knot. In contrast, many adults invoked spirits, moral transgressions, and supernatural interventions.[13] When we see young children's wonder as infused with supernatural agency, we impose the encultured interpretations of adults.

The child's wonder resembles the act of exploring a new environment. When backpacking in an alpine wilderness for the first time, as the terrain unfolds ahead, you "wonder" what is around the next bend and usually walk a bit faster to find out. You aren't expecting any one thing in particular; instead the terrain you have traversed and the surrounding scenery suggest structured possibilities: an alpine meadow with far-off views, a glade with a stream, a rocky moraine to traverse with caution. Your projections are constrained. You don't expect to see ocean surf, or a rain forest, or an amusement park. When Captain Cook rounded Cape Horn in his first voyage and sailed into the enormous Pacific, he wondered what he would discover, but his wondering had structured possibilities: a vast southern continent, a long peninsula sticking out from Asia, an impassable doldrums. He had some hints from his few European predecessors, but those only served to add gentle guidance to his wondering.

Wonder becomes more elaborated and thoughtful with each wonder-driven exploration. Expectations are fleshed out and suggest more structured possibilities. The ecologist Suzanne Simard, who discovered the fungal nets that link trees underground, first wondered about how they were interconnected and why. After learning that those nets were linked to nutritional support, she then expanded her wondering to how such connections were related to evolutionary theory and how they might pose riddles relating to theories of altruism.[14]

Our sense of wonder usually occurs in sentence frames such as "I wonder why . . . ," "I wonder how . . . ," or more generically "I wonder if . . ." or "I wonder whether . . ." It entails a tentative expectation or interpretation that wonderers are eager to confirm or discredit. It occurs not as a precise formal prediction but as an informed class of conjectures. Many of my intuitions about wonder arise from watching young children, especially our three sons and our granddaughter Frances. Frances, now at age two, has an insatiable hunger to explore and find out what is happening and why. In any new space, she immediately roams all over until she has a good sense of the layout (she is not shy). She quickly zeroes in on any new device or animal and immediately inspects it—not as the patient noninterfering invisible observer required in some anthropological field studies but rather in a headlong, grasping, groping, smelling, tweaking frenzy. She does everything she can to the new entity to learn more about it. We must initially limit these activities to save her, or the object, or both, from injury. With each successive encounter, she shows more tailored forays that acknowledge the risks associated with some objects (cats are not to be pounded).

These patterns of investigation are plentiful in all young children, as are the squeals of delight when a particularly exciting new feature is discovered. The drive to engage in activities that nourish wonder can be massive. Many times in the past few months, we have had to pull Frances away from one of these engagements to remind her that she is late for lunch, is ready for bed, or has to come inside from the snow. Her protests indicate an intellectual hunger that can trump physiological hunger, fatigue, and physical discomfort. Why should young children be endowed with such a drive when motivations relating to nutrition and shelter seem more important? Wonder, as meant here, gives a child a feeling of agency and self-efficacy. She not only learns more about the world and how things work but she learns more about how to do things to accelerate this understanding. She learns different ways of manipulating objects. She becomes skilled at accessing information in the minds of others. She discovers that the more she wonders about something and acts on that wonder, the more informative and gratifying that process becomes. Each new insight motivates more sophisticated conjectures.

Wondering builds and elaborates successive wondering. It is a perfect setup for creating an addiction to the process of discovery. When Frances first found our piano at eleven months, she was engrossed with how her pounding on keys produced sounds. Over the next several months, largely on her own, she has built on this insight to learn that parts of the key board produce high and low sounds, that individual keys make cleaner sounds than trios of adjacent ones, that pushing keys in certain orders makes especially pleasing sounds, and that some rhythmic-like patterning is involved. She is interested in how this all comes about inside the piano, but we haven't yet figured out how to let her safely explore the insides.

Our meaning of wonder is guided by observations of children and by descriptions of their behavior in the research literature. Wonder is an enabling and motivating process. It allows us to experience life more richly and imbues things with more meaning. This last point is controversial and was famously criticized in Keats's poem *Lamia*.[15] *Lamia* construes Newton's breaking down the components of light by using a prism as destroying the beauty and mystery—he "unweaves" the beautiful tapestry of the rainbow into homely brute physical components.

Mark Twain seems to echo Keats when he bemoans how serving as a pilot and learning all about how the river "works" ruined it another way:

> No, the romance and the beauty were all gone from the river. All the value any feature of it had for me now was the amount of usefulness it could furnish toward compassing the safe piloting of a steamboat. Since those days, I have pitied doctors from my heart. What does the lovely flush in a beauty's cheek mean to a doctor but a "break" that ripples above some deadly disease? Are not all her visible charms sown thick with what are to him the signs and symbols of hidden decay? Does he ever see her beauty at all, or doesn't he simply view her professionally, and comment upon her unwholesome condition all to himself? And doesn't he sometimes wonder whether he has gained most or lost most by learning his trade?[16]

Twain, however, may be describing two phenomena here. He offers the idea of deeper understanding displacing beauty, but he may be also alluding to a different effect—how intrinsic interest can become undermined, or how play can be turned into work. Motivational factors can both suppress and support spontaneous enjoyment, and Twain has returned to that theme again and again in his other writings. Keats may also sometimes intermingle how joyful play can be corrupted by work with the idea that beauty and awe are ruined by science. It is quite easy to confuse one for the other. Keats's views of science were also more nuanced than suggested by his preferences with prisms. In his short life of twenty-five years, he won science prizes in secondary school and, as a medical student, became fascinated by the nervous system and perception among other topics. Medical allusions appear throughout his writings, and, in his poem *On First Looking into Chapman's Homer*, he exalted the joy of exploration and discovery.[17] He was clearly prone to wonder, including perhaps wondering about when science elevates beauty, when it defiles beauty, and the tension between the two.

Of course, we all know pedants that can make anything boring, but those people aren't just scientists. We remember how a stirring section from a novel was destroyed by an inept "close reading" or how a fascinating event in history can become bogged down in countless caveats and footnotes. Science is no better or worse in this regard. Moreover, wonder traverses all the disciplinary terrains as the inspirational fuel driving interest and exploration. As we learn more about youthful wonder and how it can develop when everything goes well, the potential harmony between the sciences and the humanities increases.

Wonder requires some moderation and pacing. You can't continuously wonder more and more about something without pausing from time to time to digest what you have learned. In that way, wondering is like eating an especially rich dessert. I may wonder why and how sunflowers turn toward the sun and may eagerly learn various steps in the process of how differential growth rates in immature plants cause tracking movements. But at some point, perhaps at the level of how sunlight variation turns gene pathways on and off, I stop wondering about sunflowers and allow my insights to consolidate and become integrated with the rest of my knowledge.

I then can easily switch wonder to related questions (why don't roses do the same thing?) or to completely new topics. Being in a perpetual state of wonder does not mean relentlessly drilling down a reductionist hierarchy to a final molecular or subatomic account. People agilely hop to different sets of questions upon attaining some level of satisfaction, or perhaps frustration, with a particular topic. As knowledge about related but more general topics expands, such as a better understanding of gene regulation pathways, you might return to that level of explanation and resume the quest for deeper understanding of sunflowers. Revisiting and diving deeper are standard operating procedures for young children. They should be standard for adults as well.

Wonder, as used here, is a drive to explore, discover, and understand. It involves creating conjectures and entertaining possibilities. Wonder-driven exploration is cyclic with each successive episode becoming more elaborated sketches of what may lie ahead. It is a joyous, even euphoric, activity that increases feelings of agency and self-efficacy while at the same time instilling a sense of well-earned humility. You learn how to engage the world in ever more rewarding and insightful ways. A blurry grayscale sense of causal patternings becomes a clear and brightly colored awareness of why something is the way it is and how it works while also unveiling whole new terrains of wonder. Wonder imbues things with greater beauty and meaning and is at the heart of an interest in science.

Wonder is a spontaneous passion that emerges early in childhood, but, all too often, it dissipates as children grow older. When it declines, science loses its appeal and may even come to be seen as distasteful. Such declines need not be our destiny, however. At certain times and places, just the right convergences of individuals and community attitudes has resulted in flowerings of wonder long after childhood. A few brief examples illustrate what is possible.

Ages and Communities of Wonder

While wonder frequently diminishes with age, it doesn't always decline. Communities can experience widespread wonder and enchantment with science. Historians have described periods when groups of people, often from diverse backgrounds, came together in an interdisciplinary rhapsody of wonder that cut across many traditional boundaries. For Richard Holmes, the author of *The Age of Wonder*, such a period was from about 1770 to 1830, the last part of the Enlightenment. Holmes views the obsessive rational discoveries of Newton and the careful thought of Bacon as metamorphosing into more passionate forms of thinking. He describes a new period of an almost giddy euphoria in which scientists and poets (it was the Romantic period after all) inspired each other, providing metaphors and even models for the other to consider. The public was also eagerly invited in to join the discussion.

The chemist Humphry Davy was a central player during this period. Davy came from modest beginnings and was largely self-taught early on until his passion for chemistry was supported by kind mentors. In his twenties and thirties, he gave lectures to the public at the Royal Institute and the Royal Society in London. The lectures were usually open to everyone, which mostly meant a mix of the aristocracy and the middle class (few paupers attended). But for its time, it was an extraordinary opening up of science to much broader audiences of nonscientists that extended far beyond elite patrons. In reading accounts of many of these scientists as well as of their literary companions, one gets caught up in the excitement of discovery and the love of exploring. Indeed, science and geographical exploration were often intertwined when explorers of unknown parts of the world came back to the Royal Society (whose members were often their sponsors) to lecture on their trips. The childlike sense of wonder is unmistakable. The poet/critic/theologian Samuel Coleridge offers many examples in his arresting prose:

> The first man of science was he who looked into a thing, not to learn whether it furnished him with food, or shelter, or weapons, or tools, armaments, or playwiths but who sought to know it for the gratification of knowing.[18]

In 1802, Coleridge attended Davy's extensive lecture series on chemistry at the Royal Institute and justified his perfect attendance on literary grounds, remarking that he "attended Davy's lectures to renew my stock of metaphors."[19] Davy was the ideal scientist for this period because his lectures on major scientific advances were utterly captivating and immensely popular to all who attended. He fostered a vibrant and remarkably diverse community fascinated with knowing more, a community that extended, through collaborations and competitions, to France, Germany, and other parts of Europe and even to the recently created United States.

Interdisciplinary flowerings have been described at other times in history. In the West, during the Golden Age of Greece from 500 BCE to 300 BCE, Plato, Socrates, and Aristotle proposed ideas and models that are still discussed today. The "Anadalusian Enlightenment" or "El Andalus" occurred between roughly 1000 CE and 1250 CE.[20] Scientists, mathematicians, and humanists from diverse backgrounds came together in medieval Spain in a burst of creativity and scholarship. Baghdad has been called the center of a Golden Age of Islam starting at around 750 CE. In China, the Song Dynasty (960–1279 CE) is often characterized as period of remarkable intellectual freedom, creativity and innovation. A golden age has been described in central Asia (Kazakhstan, Kyrgyzstan, Tajikistan, Turkmenistan, and Uzbekistan) from roughly 800–1100 CE, with major advances across the sciences and the humanities.[21] There are surely hundreds more cases in which communities became inspired by a few charismatic lifelong "wonderers" and were protected by governments that tolerated and even encouraged new lines of thought and creative work. The European Enlightenment may have had a huge impact on history up to the present,[22] but vibrant communities of wonder seem to sprout up for brief intervals throughout the sweep of history.

My elementary school years coincided with an exuberance for science in my community. This occurred largely because of the successful launching of Sputnik 1 by the Soviet Union on October 4 1957. The ability of our bitter cold-war enemy to put something in orbit "over our heads!" aroused strong reactions in the United States, much of which centered around our failings in science education and our neglect of the next generation of scientists.[23] The United States responded with an avalanche of resources, funds, and new ways of teaching science. It is unfortunate that this surge of support for science education was fueled in part by fear and nationalism, but, within my circle of friends, the perceived Soviet threat simply added to the excitement.

From the first grade on, I witnessed a rise of science demonstrations and projects in school and a rapid growth of science projects in our homes. Virtually all my friends owned at least one chemistry kit, many of which were surprisingly poisonous and dangerous. I'm pretty sure one of my kits had arsenic and bits of radium and that my friends had other more pyrotechnic components. In addition, a never-ending stream of science stuff arrived at our house. These packages came from grandparents, aunts, and friends of the family. Rocket kits showed up at our door, not just of the vinegar and baking soda type but of much more powerful chemical mixtures. I was the proud recipient of electronic circuit kits, crystal radio assemblies, Erector sets, and The Visible Man, which was a transparent plastic human shell in which you could place all the bodily organs like a 3D jigsaw puzzle. The Man was followed later by The Visible Woman with a pop-out pregnancy unit. I received in the mail every month little blue boxes called Things of Science, which provided the materials and instructions needed to perform experiments in biology, chemistry, and physics.

Ominous portrayals emerged of how badly the Russians were beating us. The education scholar Jeffrey Herold described a CBS broadcast that reported on a "typical" Russian teenager suspiciously named Ivan who seemed to excel in all areas of study and extracurricular activities.[24] The show then interviewed teenagers from Tennessee about their thoughts on the report. The Tennesseans did not think much of Ivan, declaring him to be a bore who studied too much and probably wasn't very good at getting along with others. Herold saw it as a devastating portrayal of Americans. They cared about learning and achievement while we only cared about peers and popularity.

In my school, I frequently heard reports of how far ahead our Russian peers were in math and science and how much harder they worked. I distinctly remember statements that the best students were studying "calculus" before high school. I didn't really know what calculus was except that I knew it was not even studied in most US high schools. This glaring disparity was certainly repeated beyond my neighborhood. *The Sputnik Moment*, a documentary film by David Hoffman, details how, nationwide, we were told that Russian children went to school six days a week and were in school for a month longer each year. They averaged four hours of homework a night while we averaged thirty minutes. We were repeatedly told that Russian children had much more math, science, and engineering instruction than we did. Somehow, none of these reports about Russian achievements were daunting or intimidating to my peer group. We regarded them as challenges to be met and were confident that we could meet those challenges and have much more fun than the Russians in the process. We also somehow sensed that the Russians were probably exaggerating (see figure 1.1) and that we shouldn't worry if we weren't immediately doing calculus in the fifth grade. In retrospect, our confidence may not have been completely foolish. After all, in 1961, President Kennedy announced to the country that we were going to land on the moon, and eight years later we did just that.

A massive infusion of US funds into science and science education ensued. Everyone seemed to celebrate all that was science, math, and engineering. If you were five and had a lot of wonder, it was a pretty great time. In retrospect, I appreciate how the science-loving frenzy and the ballooning resources were uneven. Despite some attempts to include girls, science was heavily gender biased and almost completely ignored minority groups. Nor was it interdisciplinary. No one attempted creative fusions with the humanities. The only interest in fusion was thermonuclear. Yet, given my small corner of the world and my early experiences, that explosion of interest in science helped fuel my own little age of wonder. Flourishing pockets of wonder are surely happening in various places around the world today as well, but they aren't nearly as pervasive as they might be. They should be seen as the norm, not as rare, exotic, and fragile flowers.

In addition to the developmental decline in wonder, has there also been a drop in adult scientific wonder since the middle of the past century? Fond

Figure 1.1
A third-grade student in Moscow in the early 1960s, seeming to confidently answer a question in an attentive classroom. This is just one example of an avalanche of similar depictions that circulated in the United States after the embarrassing launch of Sputnik. Despite being told that Russian children were far ahead of the United States in math and science in terms of curriculum and work ethic, rather than being discouraged by such comparisons, many children of my generation were inspired by such possibilities as the United States greatly ramped up support for science education.

reminiscences of "the good old days" are all too often reflections of a recall bias rather than careful historical analyses.[25] Riding a wave of youthful science nostalgia can easily cause a mistaken generalization that the culture as a whole was more positively engaged with science. For those reasons, I must be cautious about drawing such a conclusion. That said, the world of science and engineering in the 1950s and 1960s is very different from today. As we consider developmental changes, it will be useful to occasionally consider how science and technology and the surrounding culture have changed as well over the past seventy years. In several cases ranging from educational practices to the nature of science and technology, there are reasons to worry. There is certainly at present ample discussion of the importance of training our youth for science, technology, engineering, and mathematics (STEM) careers, but that emphasis is not the same as nurturing a self-sustaining interest in science in young children and continuing to have that intrinsic interest as adults.

Wonder has social consequences. It can be rebellious and disruptive. Acts of wondering, especially asking why and how, have been called moral vices. Augustine and many others condemned the seeking of knowledge and understanding as abominations arising from lust, pride, and vanity.[26] Such sentiments often lead back to Adam and Eve's original sin at the tree of knowledge and the inference that it is wrong to want to know why and how. Engaging in such actions could lead to blasphemous challenges of religious doctrine, to atheism, and to acquiring forms of forbidden knowledge.[27] Active, free wonderers can be seen as heretics who challenge the status quo. Yet, while many have characterized questioners and knowledge seekers as prideful and vain, psychological research shows just the opposite. The more you know, the more you realize the limits of your knowledge.[28] As Socrates observed many centuries earlier, knowledge enhances humility, not arrogance. Wondering is extraordinary in that it is an act both of humility and of intellectual independence. We all have a right to wonder about why things are the way they are and how they might be different. To preserve and nourish the child's sense of wonder intact is to preserve everyone's right to question anything and to explore alternative possibilities. More mission-oriented episodes of wondering can certainly be less noble, such as wondering about better weapons during the cold war or about the possibility of improving humans through breeding during the eugenics movement in the United States,[29] but to suppress or prohibit those acts is surely not the way to devalue them. In its pure form, Rachel Carson saw wonder and humility as intertwined "wholesome emotions" that "cannot exist with a lust for destruction."[30] When wondering is not task driven and is uncensored and intrinsically motivated, we may inevitably come to appreciate the moral pitfalls laying along some paths of conjecture.

Beyond the Natural Sciences

This book focuses mostly on the rise and fall of wonder in the natural sciences and engineering. Why not all other fields of inquiry? Children and adults can clearly experience knowledge hungers and exploratory drives for the social sciences and the humanities. A child can ask why friends fight more with each other when they are both tired, why popular toys cost so much, why children's books have so many talking animals, or why most songs go up and down in notes instead of continuously up or down. These, and countless other questions, can be just as fascinating with their own paths of exploration and discovery.

Our focus on the natural sciences and engineering converges with much of the historical and current discussions of wonder and related cognitive processes. In addition, as in the Sputnik moment, sciences and engineering have once again become critical topics at national and multinational levels in ways not seen in other fields of study. Science has had arguably the strongest influence on improving the human condition. While there will always be cases in which science and technology have resulted in immense pain, suffering, and death, the overall impact of science and technology is positive and continues to move upward.[31] The cognitive scientist Steven Pinker describes how the smallpox vaccine alone, an achievement that critically depended on years of developing a science infrastructure, saved over 200 million lives in the twentieth century alone. The worst wars and genocides pale in comparison. Technology growth has helped fuel global warming, which may in the end be responsible for hundreds of millions of lives, but without science and technology, we may have never realized and understood the threat and been much less able to consider ways to solve the crisis. So, while all fields of intellectual endeavor, when done well, deserve great respect and support, they do work in different ways, with science and technology often bringing about the greatest concrete changes to our lives.

One notable attempt to spread the Sputnik fever beyond the natural sciences occurred when the US National Science Foundation funded the MACOS project (Man: A Course of Study). MACOS strove to bring psychological, anthropological, and sociological topics into the same flurry of excitement that was happening in natural science education.[32] But, as often occurs in the social sciences, MACOS was controversial as opponents inferred moral and political agendas in the curriculum. By discussing practices in some cultures that included senilicide and infanticide, divorce, and even cannibalism, MACOS was seen by some as subversive and unpatriotic and was quickly shut down. In biology, wonder may rejoice in the structure and function of eggshells. In the social sciences, wonder often has to walk on eggshells.

Some readers may not be as interested in the natural sciences as they are in topics in the social sciences and humanities. Because many of the arguments apply more broadly, we'll occasionally consider other areas of inquiry. Moreover, infants and toddlers show early skills in areas we would loosely call the social sciences and morality, and they certainly can appreciate aspects of arts and humanities, such as music. Social science or natural science "majors" don't exist in the early years. Infants and children are liberal arts students in the truest sense. They love to learn about virtually anything that they come across. Those early broad-minded versions of ourselves promise that wonder can be kept alive and nourished for a lifetime. We don't have to lose interest in the natural sciences and engineering because our dominant interests or careers move in other directions. We may choose to specialize elsewhere, but we can always appreciate the surprising and delightful ways in which insights from all disciplines enlighten and inform each other.

▪ ▪ ▪

A major theme of this book might be summarized as "Enquiring minds want to know." This phrase captures the active knowledge-seeking drive that roars to life in the preschool years. It also is the motto of the tabloid *The National Enquirer*, where the primary quest is for gossip and embarrassing details about celebrities. This contrast is important, because it reminds us that we all seek information for multiple reasons. Still, some of the original, most powerful, and most satisfying knowledge quests must also include those that expand our appreciation of the casual complexity and mechanistic beauty of all that is around us. They enable us to engage with the world more deeply and fruitfully, to see it more fully and to appreciate how lucky we are to have the capacity to expand our understandings in this way. Young children do ask questions on other topics as well, such as difficult questions about religion and sociocultural practices, although they may not find the same rewards of revealing and ever-expanding answers.[33]

To be sure, we also sometimes eagerly scoop up and spread gossip, almost invariably in forms that harm others, but this rarely feels good and, over time, feels worse and worse. Gossip doesn't exist in young children or infants. Their delight in actively acquiring new information comes from a deep human desire to understand all that is around them and to share those experiences with others. This desire is clearly not the only drive that motivates human behavior. We also have appetites for social companions, for experiencing pleasure and even some forms of pain,[34] and for learning social networks and our standing in them. Just as many animals have specific hungers that increase when one nutrient, such as protein or glucose, becomes too low,[35] humans may have different mental hungers that need to be satisfied. Wonder may be one of the most enabling yet oft neglected of such hungers.

Wonder is more than mere curiosity and goes beyond seeking brute facts or collections of trivia. Wonder isn't implanted in the infant's mind. It is not instilled or encultured. It is a universal, native quest for understanding fueled by an exploratory drive that provides important adaptive benefits to a growing human. Wonder is revealed and supported by the extraordinary ways in which infants and young children grasp and expand on their causal understandings of what is around them. It is also deeply social, flourishing in conversations, pedagogy, and interdependent knowledge networks.

A closer look at the goals of wonder and the resulting internal mental representations reveals a particular mode of thought as central: the mechanistic mind, that is, a deep interest in how things work. Children appreciate early on the special powers of mechanistic explanations and seek them out, share them, and deploy them with great agility. They seek out mechanistic details but then use those details to create abstract causal generalizations about different classes of phenomena. Those more general causal sketches help them understand new phenomena even as they may forget the lower level details that built those sketches.

While wonder may arise spontaneously in all young children, it needs to be nourished and supported by others to flower. This can occur early on primarily through appropriate interactions with others. However, it can also be discouraged and smothered, and this happens all too often. It happens among even the most well-intentioned adults if they misconstrue how children naturally understand the world and how their understanding changes with development. It can occur when adults and the culture at large create systems and processes that undermine the motivations that support wonder-driven activities.

When wonder is stifled and demotivated, it can lead to cognitive decay that makes us all more susceptible to misleading cognitive biases, misinformation, and the blind following of consensus. Ultimately, a life without wonder can lead to disengagement, disillusionment, and even distrust of science. When this happens, we lose touch with the ability to engage in meaningful discourse about science when it matters most.

Fortunately, even when everything has worked against wonder, we can also reawaken it in ourselves and in our communities. This isn't just blind Pollyannaish optimism. It calls for a program of small- and large-scale activities, mostly low-cost and easily implementable. These activities can rekindle the flame inside of us and allow us all to share the glowing warmth of wonder. We therefore start with the youngest humans.

2 Early Exploration and Discovery

Young children don't always get things right. They can develop ideas about nature and technology that most adults see as clearly wrong. But, just as even the most advanced scientists sometimes also get things wrong, it is how they are mistaken that matters. What does their error tell us about how they think and how they build their knowledge and understanding? In many cases, impressive inferential machinery is revealed. One, summer when our son Derek was almost four, he asked a surprising question: "Do trees have hearts?" Rather than simply say they didn't, I asked him why he asked. A flood of statements ensued: "Because people have hearts and yesterday you said Daisy (our dog) needed her heart medicine. So people and dogs have hearts. And when I asked about Daisy, you said all animals need hearts 'cause they pump blood all over and that's how food gets to all parts of their body. . . . so trees must have hearts too." When I asked why trees had to have hearts, he said, "They were alive 'cause they grew bigger, made new leaves every year and made seeds for more trees and even fixed themselves when they got hurt. So . . . they must have blood and a heart to move all their food around to do those things; maybe their blood was like maple syrup." (Some neighbors had maple trees with collection pails.) At this point, I finally said trees didn't really have hearts, and my son immediately demanded to know how they pushed all their blood around. I madly tried thinking about osmotic pressure, capillary action, and photosynthesis in leaves and realized I had nothing to offer that made any sense. So . . . I punted with the usual: "It's complicated."

A great deal of thinking and reasoning were going on here, much of which unfolds in the next several chapters. To anticipate, young children are infused with spontaneous curiosity and a drive to find certain kinds of answers. They constantly mention causal relations and make inferences that transcend what it immediately present. The developmental psychologist Jerome Bruner famously described the phenomenon as follows: "Being able to 'go beyond the information' given to 'figure things out' is one of the few untarnishable joys of life." In Derek's query, we see the opportunistic use of information that I had provided earlier in the same conversation. This ended up creating errors, but errors that revealed insight as well. How does all this cognitive machinery get off the ground? It all starts in infancy.

Well before they are able to produce or understand language, human infants are curious about how and why things happen. They rapidly develop intuitions about the causal patterns around them. Because of these early abilities, preverbal infants are sometimes described as "scientists in the crib," but such accounts quickly run into arguments about what it means to be a scientist. In many senses of the word, infants are not scientists. They do not propose formal hypotheses and test them in carefully controlled ways, and they do not participate in arguments concerning their views. Yet they do have interpretative and predictive skills similar to those used at all ages to make sense of complex systems. Infants also explore objects in ways suggestive of "hypotheses." These preverbal exploratory activities are early signs of their drive to learn. Both in terms of tracking causal patterns in the world and in terms of using strategies for learning more about the world, infants can be surprisingly proficient.

The past few decades of research on infants' and children's minds consists of many thousands of studies covering a huge range of topics.[1] A one-line summary of much of that research is a string of discoveries showing far more cognitive ability in infants and young children than was previously believed. These early cognitive capacities provide critical components for a developing sense of wonder. To see how this occurs, we need to clarify what aspects of thought are needed to grow the earliest traces of wonder into fully functioning engines of imagination and discovery.

Four components greatly facilitate an emerging sense of wonder in the individual child: (1) a rapid and automatic way of gathering data about the world, (2) an appreciation of the distinctive role of causal relations, (3) an ability to conceive of domains of things that cohere because they share common causal patterns that form stable clusters, and (4) a sense of how such domains might interact in systematic and predictable ways with other domains. Collectively, these skills support a cycle of activity that enables children to puzzle over a problem, to formulate conjectures or hypotheses about possible alternatives, and to consider counterfactuals before updating their puzzling. These four abilities and the cycle undergird wonder and allow it to be an engine of conceptual change. They do not exhaust aspects of cognition that support wonder. Others include executive functioning, the ability to simultaneously represent several alternative models or theories, logical and analogical reasoning, and an ability to weigh probabilities, but I see these four as the central ones that arise in infancy.

This discussion of cognitive development of a child as an independent learner will then enable us to consider in chapter 3 how wonder goes beyond the individual mind to expand greatly in power by drawing on what others know. Finally, the most powerful and relevant use of wonder for science involves a special appreciation for what we describe in chapter 4 as the mechanistic mode of understanding.

Data Aggregators

To wonder about the world requires at least a crude sense of stable patterns in that world. These patterns recur often enough to be noticed among the noise and confusion that also confront the child. Science relies on the tabulation of statistical patterns in time and space. When certain events co-occur repeatedly, such as lightning and thunder, we notice and remember that relationship. Scientists track correlations, frequencies, and contingencies and use those patterns to make predictions and support explanations. Every adult automatically engages in this tracking of correlations and contingencies at an implicit level all the time.[2] Infants are also impressive statistical learners. They notice and remember correlations and contingencies long before they have any way of explicitly talking about them.

Infants' abilities to track patterns were demonstrated in a series of landmark studies that first appeared in 1996.[3] The studies focused on how infants identified artificial spoken words embedded in much larger strings of spoken syllables. When infants hear a string of syllables, such as po-ta-to, repeatedly occurring, do they start to see that string as somehow special and cohering as a unit, namely *potato?* They do so easily. In fact, infants under six months can learn new artificial words in under two minutes. (They are artificial to make sure infants don't already have experience with them as words.) In the studies, infants listen to much longer strings of syllables in which certain artificial words, such as *bidaku*, repeatedly occur amid other nonrepeating random strings of syllables. No other cues to repeating triplets are present such as intonation, special stress, or pauses. Based solely on repeated co-occurrences, infants reliably came to expect such sequences to occur again in the future in comparison to other completely novel three-syllable sequences for which they showed no such expectations. Because novel word learning can happen when the co-occurrence patterns are highly likely but not perfect, we can think of the infants as mentally tabulating things like correlations and probabilities.

This kind of early learning has been called *statistical learning*, and young infants display such an ability not just for repeating speech sounds but also for musical tones and for repeating visual patterns.[4] Infants and adults learn these statistical regularities without any awareness that the learning is occurring. Statistical learning can occur automatically without effort or goals to learn.[5] It may therefore also be involved in creating implicit attitudes,[6] those sets of impressions we develop toward others and that have been linked to racism, sexism, and other antigroup impressions.

The ease with which young infants do automatically learn statistical patterns cannot explain many aspects of learning,[7] but it is an important part of a cognitive system capable of wonder. We need to constantly update our mental database of the recurring patterns that we encounter. To wonder why cats crouch before they leap, you have to first notice and remember that regularity. But humans go far beyond noting co-occurrence patterns. We overlay those patterns with causal interpretations that provide new meanings to events and that impel us to learn more about them.

The Causal Connection

A fundamental skill in understanding the world involves detecting causation as opposed to mere correlation. This contrast is stressed even at the university level as faculty frequently caution their students not to leap from strong correlations to conclusions about causal relations. For example, the positive correlation between years of education and health does not lead to the simple conclusion that more education causes better health. There are likely causal links between the two measures but they run in both directions and often involve interactions with other factors such as income.[8] A more obviously silly faux causal link between two variables is the strong correlation between toenail size and the syntactic complexity of a child's sentences. No one thinks toenail size directly influences syntax or the reverse, but it is easy to see how a third variable related to maturation and growth could causally influence both and create the correlation. Spurious correlations are even worse as they have no link to any feasible causal patterns. These can be found by inspecting thousands of data sets and by slicing up the data into bins of just the right size. With increasing ease of scanning such data sets, a cottage industry of correlational humor has emerged. Spurious large correlations have been found between funding for science in the United States and death by suffocation and hanging, between divorce rates in Maine and consumption of margarine, and between the number of letters in the winning word of the national spelling bee and the number of people killed by venomous spiders.[9] When shown these cases that are so blatantly at odds with any feasible causal account, all adults will readily agree that correlation does not always mean causation.[10]

Infants cannot explicitly describe the contrast between correlation and causation, but more tacitly they show behaviors indicating that they are often tuned to the same sorts of patterns that adults see as causal and not merely correlational. Imagine observing three events involving both a black and a gray billiard ball (see figure 2.1).

In one event (figure 2.1a), the black ball rolls toward the gray one, and, right after it touches gray one, the gray ball quickly moves off as the black one suddenly slows down. This is commonly described as a "launching event" where the black ball "launches" the gray one.[11] In contrast, another event (figure 2.1b) inserts a one-second delay before the gray ball moves off, or the black ball might stop suddenly four inches away from the gray one, which nonetheless moves off after the black one stops (figure 2.1c). In either of these two instances, the motion of the black ball is not seen as causing motion of the gray one even if the event is repeated several times and thereby demonstrates a reliable correlation. Using experimental designs in which videos show similar event contrasts, researchers have repeatedly demonstrated that infants reject some reliable correlations as implying causation.

Classic experiments by the psychologist Alan Leslie in the 1980s showed that infants see "launching" in the same way as adults.[12] Several decades of follow-up studies have convincingly supported these early findings, and, while current researchers still debate over whether such causal impressions are present at birth or take time to learn in the first few months, virtually

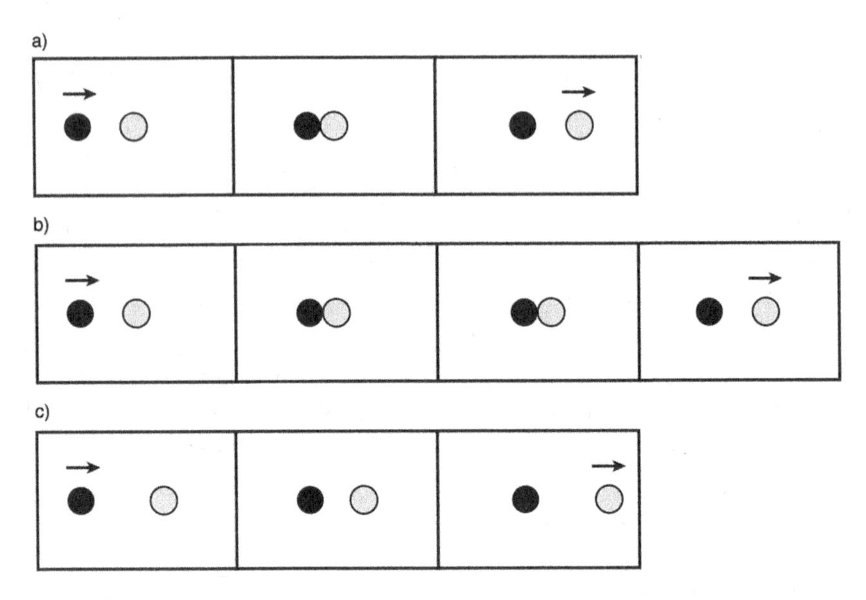

Figure 2.1
Perceiving causality. Panel a shows a simple case of launching. Events of this type elicit a strong impression of the black ball causing the movement of the gray one. In panel b, a one-second delay occurs before the gray ball moves. In panel c, there is a spatial gap between where the black ball stops and the gray one starts. Both manipulations block the sense of a causal interaction between the two balls.

all contemporary scholars agree that well before their first birthday, infants show a special sensitivity to causal patterns by noticing them more quickly and remembering them more strongly.

Infants also start to distinguish among types of causality. For example, while adults describe some interactions between two objects as "launching," they see others as "triggering." When the first object provides all the energy needed to set the other one in motion, it is launching, but when the first object releases stored energy within the second object such that it moves off at a faster speed than could be possibly transferred from the first object, they see triggering. Infants at eight months make the same contrast, as shown by members of our lab.[13] Triggering is analogous to a classic mousetrap. The highly sensitive catch plate need only to be touched lightly by the mouse for it to trigger the release of all the energy in the spring that brings the bar known as the "hammer" down on the unlucky mouse.

While researchers still actively debate when the first causal intuitions emerge in infancy and how they are represented in the mind, they all agree that, by twelve months, infants are especially attentive to causal relations. Their tendency to notice and remember causal relations also forms the basis for their ability to notice those clusters of causal relations that distinguish different domains of things.

Demarcating Domains

Causal relations often come in characteristic groups or clusters. For example, in thinking about two people, I might notice that, in contrast to a pair of billiard balls, they often causally act on each other at a distance. A remark to a person several feet away can cause that person to move quickly backward. Cause-and-effect relations for people have longer time lags than for balls. You do not move instantaneously after I speak, unlike cases in which one ball launches another. There is a noticeable lag. People move on their own without needing any external force. Simple balls do not spontaneously move. Self-generated motion conveys the strong impression that something inside the mover is causing the movement. People can move in irregular ways, darting this way and that. Balls move in smooth predictable paths unless something else intervenes. People interact contingently; balls do not. There is a back-and-forth rhythm to many human social interactions whether they be conversations, silent greetings, or hot pursuit. Taken together, several interacting causal relations distinguish the motions of people from those of simple solids. And of course, these patterns extend far beyond humans to dogs and cats and all sorts of other less familiar creatures.

By twelve months, infants are well aware of these two contrasting pairs of causal regularities related to motion.[14] They may have not yet integrated all the relations that toddlers notice, but they notice a complex set of interactions that give them a sense of two starkly contrasting realms, one concerning agents and one concerning simple solid objects. Infants' awareness of these two realms, or domains, is sometimes described as the earliest forms of an intuitive psychology and an intuitive physical mechanics. But an intuitive psychology, no matter how primitive, ought to include some notion of internal mental states. For that reason, a great deal of research has focused on whether infants also make inferences about goals.

Reaching is an action that automatically suggests goals to adult observers, but not just any goal comes to mind. When you observe someone reaching for an object, you normally assume they have the goal of contacting that object and probably retrieving or manipulating it. You don't assume they have the goal of moving their hand to that specific location in space without any interest in the object itself. Do infants have similar assumptions?

An influential line of work initiated by the psychologist Amanda Woodward in 1998 has examined inferences made by infants when they observe a hand reaching for a particular object. For example, a hand might reach for a teddy bear on a pedestal at a specific location A.[15] This reaching happens repeatedly until the infants no longer find it interesting. The infants are then shown one of two displays: (1) The hand moves to the same location A but now is moving toward a different object such as a ball, or (2) The hand moved to a new location B but toward the same teddy bear as before. In this pair of situations, infants look longer at reaches for the ball in the same location than for the teddy bear in a new location. Looking longer is interpreted as evidence that the infants' expectations are violated, or in more familiar terms, they are surprised. (They do not, however, usually show facial expressions of surprise that we so commonly see in older children and adults.)

Now consider the same set of events with the teddy bear and the ball, but where, instead of a hand, the initial repeated action is executed by the movements of a clearly inanimate wooden stick. As adults, when we watch a stick repeatedly move toward the teddy bear in location A, we develop an expectation that it will again go to location A when the ball has been moved there and the teddy bear is at location B. We assume that the stick just mechanically and deterministically follows the same path as always, and we definitely do not attribute any goals to the stick. Infants as young as five months have similar expectations. They look longer when they observe the stick "reach" for the old object in the new location. They too see the stick as destined to move in precisely the same physical path again and again. The hand's movements are seen as being controlled by an actor with the goal of obtaining the object regardless of its location.

Infants quickly code objects as either goal-directed or not by using several kinds of cues. The object might look like a hand or contain a face-like pattern. It might respond contingently to the actions of another object some distance away.[16] It might move in an obviously self-propelled manner. Even a simple triangle can acquire goal directedness if contingent changes of movements create an impression of the triangle "chasing" another geometric figure.[17]

As adults, when we see self-propelled objects with goals, we also assume that the causes behind all these movements come from within the objects themselves. We see humans and animals as acting in ways that arise from events happening inside them and do not make such inferences for simple inanimate things. Our research group initiated a series of studies testing this idea with fourteen-month-old infants. The infants watched videos in which animated cats either swayed back and forth or jumped up and down.[18] The cats were unusual in that their insides were visible through a kind of semitransparent skin and fur, a manipulation that was surprisingly unproblematic for our infant observers. Each cat was depicted with hats and stomachs of the same color, red or blue. When shown new videos in which the stomachs and hats of each cat differed in colors, infants expected cats with the same colored stomachs to move the same way and ignored whether or not the hat was the same color across movements. Hat color was seen as irrelevant to a cat's manner of motion while stomach color was seen as critical. Using different measures and different versions of agents and inanimate counterparts with revealed insides, other groups have found the same expectations about insides in infants as young as ten months.[19]

This preference for insides in understanding animate creatures may be a precursor sign of a more general bias to infer that surface properties and behaviors of certain kinds of things (e.g., animals) arise causally from essential inner features. An early version of such an *essentialist bias* could provide a useful strategy for guiding exploration of a thing given its general category. If something seems to be an animal, assume that the most important things to know about it are inside.

Infants know more than the contrast between an early physics of simple inanimate objects and a psychology of social things, although perhaps in fragmentary forms. Two other examples from infants' understandings of plants and liquids make the point:

Infants apparently view plants as both important and potentially dangerous.[20] They are more reluctant to touch novel plants than novel control objects that are superficially shaped and colored like plants but clearly appear as human made. This reluctance is equally strong for plants with delicate soft leaves and plants with sharp thorns, suggesting a deep-seated sense that plants of all kinds might mean trouble (as irritants or poisons). At the same time, infants carefully observe how other humans behave around plants and take special notice of any plants that adults bring to their mouths in comparison to the artifacts they bring to their mouths. Plants are things to carefully monitor in terms of how others interact with them. They could cause great harm or they could be tasty food. Not until the preschool years and later, however, do children appreciate important commonalities between plants and animals.

Infants have distinct and specific expectations about the behaviors of liquids as opposed to solids. Thus, they expect liquids in a glass to stay horizontal when the glass is tipped but expect solids to tip with the glass. They know that liquids can pass through grids while solids cannot. They also easily extend their inferences about liquids to loose granular substances like sand.[21] They need to appreciate these contrasting causal clusters if they are to later wonder why cold transforms water from a liquid into a solid. Such a transformation must seem especially miraculous when first encountered if the observer appreciates all the causal patterns that distinguish solids and liquids.

Interactions between Domains

Intuitive expectations about causal clusters associated with broad domains such as goal-directed agents and physical objects allow us to think about how two or more domains might causally interact. How do interactions between goal-directed agents and physical objects differ from those between two physical objects? A key contrast concerns the ability of goal-directed agents to create order out of randomness. When you witness a transformation of disorder into order, you immediately infer that an agent has been involved.

These linkages of order with goal-directed agents motivated the "argument from design." Prior to Darwin, an invisible deity was often inferred as necessarily behind much of the order present in the biological world. Everyone agreed that humans could not have created most of the systematic structure that was so apparent in plants and animals, much of which clearly served obvious functions. In 1802, the Reverend William Paley argued that some kind of powerful deity must have been responsible for the highly ingeniously designed properties of the eye.[22] For Paley, the eye virtually shouted that it was created by a goal-directed agent in much the same manner as a clock.

The degree of order does not have to be nearly as complex as an eye to elicit a sense of agent involvement. Hikers routinely encounter simple clusters of rocks on beaches and mountain trails indicating that an agent must have been there. An enormous range of configurations can irresistibly suggest order. For example, four rocks strongly imply placement by a human agent if they form a straight line; form a neat square; are stacked on top of each other; or are ordered according to size, darkness, or smoothness. These examples hardly exhaust possible cues, which in turn proliferate when larger numbers of rocks are involved. This ability to detect order applies to many configurations that we have never seen or even imagined before.

Figure 2.2 shows the ground in the wild interior of Iceland. It is immediately obvious that aspects of this terrain were created by humans and did not occur naturally. The stack of rocks in the foreground and the smoothly

Figure 2.2
This scene, composed entirely of naturally occurring objects, nonetheless immediately indicates a kind of order that we would only attribute to goal-directed agents. It "insists" that an intentional agent must have created the array and that it couldn't have possibly arisen though random processes.

curving darker colored path of constant width suggest that people caused both the path and pile. Artists such as Andy Galsworthy have beautifully exploited these intuitions by finding a natural scene and then creating some kind of order that irresistibly implies the presence of an intentional agent. Given that there are seemingly infinite ways of indicating obvious order, it might seem that the ability to detect such departures from randomness would only slowly accumulate over a lifetime of experiences. Yet, the immediacy of the impression suggests there might be much earlier origins of such intuitions.

Our research group explored expectations about the appearances of an ordered world at different ages.[23] Our first studies involved adults and preschoolers. We showed them a scene of ordered things (such as a neatly organized room) and then a scene of disordered things (such as an extremely messy room with toys and furniture randomly scattered about). Both adults and preschoolers universally agreed that the change from order to disorder could have been caused by either an intentional agent (e.g., a willful but destructive teenager) or by a simple agent without intentions, such as an extremely strong gust of wind that rushed through an open window. When the sequence was reversed, however, in which a disordered scene became orderly, both age groups all agreed that such a change could only have been caused by an intentional agent. Despite this unanimous agreement, not a single preschooler could justify their strong convictions by explaining why only the intentional agent could create order. Indeed, many adults had great difficulty articulating a clear reason.

Given the strong, but difficult to justify, expectations found in preschoolers, we thought it possible that even infants might show similar expectations. We created a video of moving balls that gave cues of being either goal-directed agents or simple objects. Thus, the goal-directed-agent ball moved on its own and changed speed, had features suggesting a face, and seemed to look at the observer. The physical-object ball had the same features scrambled and not suggesting a face, entered the scene already in motion at a constant speed, and showed no indications of interacting with the observer. We then presented twelve-month-olds with displays of neatly ordered blocks that were temporarily occluded by a screen and then shown to be in disarray when the screen was removed. The infants (and adults) expected both the animate and inanimate agents to be possible causes. When the reverse event occurred, namely disordered blocks becoming ordered, infants instead expected just the animate ball to be the agent.

Using very different methods and stimuli, other researchers have found similar inferences of links between intentional agents and physical order.[24] For example, in one study, nine-month-olds expected that a hand could select nonrandom orders of balls out of a container of different colored balls but that a claw could not.[25] They sensed now that color-based ordering of a randomly distributed set of things could only be done by a humanlike agent.

We do not know how preverbal infants detect so many different departures from randomness and what forms of order might only be apparent at later ages. Nor do we know how they come to so immediately conclude that changes to more ordered states require goal-directed agents. Conceivably, this ability was invaluable to earlier generations of humans as a way of sensing when other goal-directed agents, including animals, were nearby. I eagerly await future studies unpacking how all this can cognitively occur in the mind of preverbal infants.

These examples just touch upon hundreds of studies showing that preverbal infants can think about how different domains can interact. They not only partition the world into domains (such as the social and the nonsocial) governed by different rules and causal patterns but they then also consider how those domains might interact. This is a more sophisticated form of wonder than simply exploring patterns within one domain at a time.

The PHED Cycle—The Puzzling, Hypothesizing, Exploring, and Discovering Cycle

Infants' abilities to construe the physical and social worlds in different causal terms has been described by the psychologist Elizabeth Spelke as "core knowledge,"[26] a deep-seated core of beliefs about domains such as physics and psychology that forms the foundation for later learning, especially when interactions between those domains are considered. Core knowledge is often considered to be present so early in development as to suggest that newborn infants are "prepared" to easily learn those particular domains. If that kind of knowledge is indeed foundational, what happens when infants encounter a device that apparently violates one of its principles? How does witnessing a violation influence an infant's behavior? In terms of wonder, such experiences may greatly accelerate the development of wonder through engaging in the PHED cycle. Using the cycle involves posing *P*uzzles, inspiring conjectures and *H*ypotheses, motivating *E*xploration, and affording *D*iscoveries, which can then cycle back to help pose new puzzles. This tightly interconnected combination of cognitive activities, or PHED cycle, continuously creates more and more elaborated versions of wonder while also "feeding" the hunger for new insights.

Puzzlement is one sign of responding to violations and anomalies. Psychologists Aimee Stahl and Lisa Feigenson asked how infants' explorations of objects might vary when those objects either behaved in ways that violated core physics or did not. Violations included passing through another physical object or seeming to go behind one screen and emerge from behind another without traversing the space in between. When eleven-month-olds witness such violations, they explored the deviant objects more fully and learned more from those explorations.[27] Their curiosity was triggered, and their subsequent investigations suggested that they wanted to know why the object was behaving oddly. In such cases, a combination of special cognitive states seems to occur that were described as surprise, enhanced interest, curiosity, and some version of answer seeking or exploration. Presenting children with such core anomalies can result in their being fully engrossed in the PHED cycle. By better remembering new information about objects that violated core knowledge, the infants apparently assumed it was important to specifically focus on the peculiar objects' properties so as to learn what might explain their bizarre behaviors. Similarly, when given a choice between exploring the core-knowledge-violating object and another control object, infants strongly preferred to examine the one that had just behaved oddly.

More subtly, preschoolers show increased exploration when "evidence is confounded." When presented with a jack-in-the-box toy that prior evidence suggests is triggered by either one of two handles (confounded), they will explore the object more to determine the real cause.[28] No one has to motivate the child to explore more; the child does so because of an intrinsic interest. Children recognize a gap in their knowledge, or an uncertainty about which of two possible handles made the toy pop up. Then, on their own, they seek to find out which handle is the causally relevant one.

Recognizing gaps in one's knowledge can be a challenge for even the most sophisticated scientists, but at least in some simplified tasks, young children not only recognize the gap, they seek to remedy it. Exactly how they construe the gap remains unclear. Is it seen as a missing bit of knowing how to operate the toy or as a missing bit of understanding how the toy works, or both? Such contrasts matter greatly and may ultimately be at the heart of the kind of wonder that fosters engagement with science.

Infants and young children therefore show curiosity and exploration for at least two reasons. Violations of core theories trigger surprise and exploration in infants, and, by the early preschool years, ambiguity about causes prompts exploration to determine which factor is the true cause. But these two processes, impressive as they are, do not fully characterize why children and adults engage in inquiry about their natural and engineered worlds. Consider why you might choose to read a new article in the science section of the *New York Times*. Sometimes the title may indeed refer to a provocative anomaly: "The Loudest Bird has a Song Like a Pile Driver." Other times the title may examine two possible causal patterns: "Meat's Bad for You! No, It's Not! How Experts See Different Things in the Data." But we often are simply told a fact that seems worth reading more about, such as "This Fungus Mutates. That's Good News if You Like Cheese," or "In the Sea, Not All Plastic Lasts Forever," or "The Twitch That Helps Your Intestines Grow." We are presented with the opportunity to learn more about something not because it is especially surprising or because we have clear alternatives in mind, but rather because we see the promise of an interesting story that will enable us to understand some corner of the world more clearly.

Infants and young children may also pursue information for other reasons than explaining anomalies or choosing among two clear alternatives. They may simply want to gain a better sense of what is going on around them so that they can act on their world more effectively in terms of navigating spaces, fixing objects, and predicting outcomes. We may seek out information to fill out the picture, complete the story, or finish a project. We care about closure because it provides an opportunity to pause and encode in memory what happened.

Consider how some people, including some young children, enjoy doing jigsaw puzzles for hours. There are no big surprises. (Unless the last piece has been hidden by a sly sibling.) The players know exactly what the end result will look like and are rarely startled midway. But there can still be great satisfaction upon completion. When listening to an interesting story, we can be deeply disappointed if we don't get to hear the ending. When exploring a new place such as a house or a park, we often want to get a clear sense of the overall layout and can be frustrated if we don't know how different locations fit together. Infants, like all of us, strive to achieve closure. They have a need to identify *all* the toys that are in a box, or explore *every* room of a new space. They have similar needs for closure in their causal models of how things happen. This constellation of needs for closure often helps drive wonder, especially when there is a need to close an explanatory gap.

Causal Elaboration and Integration beyond Infancy

Causal understanding grows greatly between two and five years of age. The most consistent overall developmental change during this period is increasing integration of all the different streams of causal information.[29] Children become more adept at inferring how causes are stronger when causes and effects co-occur more frequently. Infants also expand their ability to learn hidden causal relations by observing covariation patterns, namely how often one event (such as a solid object dropping down toward water) is followed by another (such as a splash). They also can learn a great deal by manipulating parts of a device. For example, they can learn whether the underlying causal process is a simple linear chain of causes and effects or a branching structure where one cause has multiple effects.[30] If infants learn that turning the crank several times on a jack-in-the-box is always followed by both a clown popping up and by the playing of music, they are likely to infer that turning the crank is an initial cause that branches into two different effects.

Preschoolers begin to entertain counterfactuals in their attempts to figure out causal relations. Thus, the infant learning about the jack-in-the-box might also realize that there could be a box where the crank only causes the clown to pop up or a different one where the crank only causes the tune. Counterfactual thought is traditionally measured by the comprehension and use of such phrases as "If x didn't happen, then y wouldn't have happened either"—for example, "If the contractor hadn't used inferior materials, the bridge wouldn't have collapsed." Young children have difficulties reasoning about such sentences but they fare better when observing visual animations. To demonstrate this effect, several labs, including ours, worked together on a project showing young children sets of brief videos.[31] Children observed computer animations in which balls moved toward soccer-like goals either in straight lines or by bouncing off brick-like walls. By at least age five, they mentally simulated different scenarios that didn't happen so as to better explain what actually did happen. They were reasoning counterfactually without having to utter a word. A convergence of recent studies shows that at least by age four, children spontaneously engage in these forms of conjectures or informal hypotheses that are so central to developing more sophisticated forms of wonder.[32]

Before they enter formal schooling, children are able to quickly deploy general causal reasoning strategies while also incorporating the peculiarities of each domain. They realize that cause-effect time lags are usually longer for psychological events than biological ones. People take longer to laugh when told a joke than a window takes to break when hit by a stone. Their "causal calculus" has become both more powerful as a general skill set and more sensitive to the marked contrasts in causal patterns that occur between broad domains, domains that often resemble academic disciplines, such as psychology, physics, biology, and economics. Wonder often is the major force behind the growth of understanding and discernment in an ever-expanding appreciation of domains and demarcation of smaller areas within those broad domains, such as biology and its different realms of animals, plants, and microbes and their many further subdivisions.

Developing Biological Thought in Preschool and Later Years

Throughout preschool, children elaborate on their early models of the world as their causal reasoning becomes more intricate and far-reaching. They are able to discern new domains that cohere because of new insights into underlying causal patterns linking together entities and processes in each domain. Biological thought nicely illustrates these kinds of conceptual growth. We briefly consider here the extensive body of research on biological thought as an example of how interpretations of the world can grow between preschool and the early school years. This growth also illustrates the potential for wonder to continue on the same trajectory for a lifetime if it is embraced and supported.

A central question in early intuitive biology concerns how children come to appreciate that both plants and animals are part of a larger category of living things. This might seem to be a highly sophisticated insight requiring knowledge of molecular processes such as gene transcription, respiration, and chemical energy storage and use. However, adults have long shown an appreciation of the overarching category of living things and have done so without any knowledge of molecular processes. What kinds of causal interpretative system were they using instead? One common and widespread system was vitalism.

In Western, Eastern, and African traditions, vitalism has been invoked for centuries and still influences informal biological intuitions in many adults.[33] Vitalism usually implies a unique force, or spirit, in living organisms. The life force enables organisms to take in and store energy (through food and water) and to use that energy to grow, recover from injury and reproduce. Vital forces also enable animals to move. Vitalism itself evolved in more recent times as a reaction against Descartes's proposal that animals were merely machines of greater complexity than those made by humans. The human "soul" had nothing to do with the life-bearing capacity shared by all animals. In reaction to these mechanistic views, scholars such as the embryologist Hans Driesch attempted to "revive" vitalism by proposing that an inherent substance in organisms allowed them to carry out more microscopic organic processes.[34] In the end, however, vitalism is no longer taken seriously in the biological sciences.

The first systematic demonstrations of widespread vitalism in preschoolers emerged from years of research conducted in Japan by the psychologists Giyoo Hatano and Kayoko Inagaki.[35] They showed that, by age five, children viewed animals and plants as sharing several common properties such as growth and resistance to disease. Moreover, the children interpreted those shared properties in vitalistic terms. When asked to choose between different forms of explanations of biological processes (such as why we eat), they preferred vitalistic reasons ("Because our stomach takes in vital power from the food"). They readily extended these ideas to other plants and animals but not to nonliving entities. These pioneering studies were later replicated in other cultures such as that of the United States.[36] Importantly, properties such as growth and taking in food and water were seen as causally related, suggesting a pattern of coherent causal beliefs that served to demarcate living things.

The full developmental trajectory of biological thought remains an active area of investigation. On the basis of the ease with which even infants contrast the social world with the physical world, the psychologist Susan Carey proposed that the infant and young child did not appreciate the category of living things and instead assimilated living things such as people and many animals into the domain of psychological agents and assimilated plants, and some "lower" animals such as worms, into the realm of inanimate things.[37] For Carey, a true awareness of the category of living kinds did not emerge until the school years when a sense of legitimate biological mechanisms emerged through formal instruction, but Hatano and Inagaki and several other research groups, including our own, proposed that children could still sense the category of living things through alternative models such as vitalism.[38]

What then develops during the early school years with respect to biological thought? One repeated finding is that earlier causal belief systems such as vitalism are rarely discarded completely in everyday thought.[39] We can clearly see elements of vitalism in adults who also have sophisticated grasps of biological mechanisms. A doctor describing the death of patient might end a discussion of physiological breakdown with the vitalistic phrase "and finally the life went out of him." The coexistence of several belief systems in one mind, many of which seem to contradict each other, remains a topic of great interest in cognitive science. Moreover, the early belief systems do not automatically block or impede the growth of new ones, even when they are mutually incompatible.

In the early school years, children develop better senses of causal patterns in biology. They remain clueless about many details, as do most adults, but as they start to wonder about how things work, children start to rule out certain classes of explanations. For example, my colleague (and wife) Kristi Lockhart led a project showing developmental changes in concepts of healing agents.[40] In one set of studies, while younger schoolchildren saw ingested medicines as going exclusively to diseased areas, older children saw medicines as going everywhere in the body. As they gained a better sense of digestion and blood circulation, they started to see how wide dispersion of ingested materials was more likely. Children continue to notice new mechanistic details that help them to refine their causal interpretations of biology. Sometimes, these conceptual changes are incremental, such as learning that insects have a blood-like fluid and a heart-like pump that moves the fluid around, but that no blood vessels are needed. Other times, the changes can be more revolutionary, such as ultimately rejecting the idea of vital forces altogether in favor of physiological systems and cycles. Throughout both kinds of change, children strive for coherence and consistency. Indeed, when an inconsistency emerges, it often provokes wonder. Why don't insects need blood vessels? If an adult explains it's because their bodies are so small, the response may elicit further wondering as to why size matters. These exchanges may continue until perhaps ultimately ending in an explanation that insect blood isn't red because it doesn't need a red-colored part (hemoglobin) to carry oxygen since it can get all the oxygen it needs directly through its "skin." (This at least was how things went in a later conversation with one of our sons.)

Our brief overview of folk biology brings us back to the discussion I had with Derek about trees. We see how, as he approached his fourth birthday, he was striving to find ways to assemble causally coherent models that could unify plants and animals. His particular idea of granting hearts to plants made sense as an attempt to find a common causal system of explanation. It wasn't correct, but it was definitely the right kind of process for growing knowledge. It was fueled by wonder. Perhaps through vitalism, he saw commonalities between plants and animals, and he had then learned about how tree sap helped the tree thrive and grow. This created an explanatory gap or puzzle, leading him to conjecture about the presence of hearts in trees. Through additional exploration, first with me as a source, and later through other sources and looking more closely at plants, he refined and reorganized his conceptual system into a form much closer to that of adults.

■ ■ ■

Through growing facility with the four elements of wonder—(1) effortlessly aggregating data, (2) appreciating causal relations, (3) valuing causal clusters, and (4) discerning domain interactions—and through engaging in the PHED cycle, infants start off with a rudimentary "wonder lite" that is the basis for the system that is exuberantly driven to learn more about the world. By the time children enter school for the first time, this system of wonder is running at full throttle and is providing rapidly expanding insights into everything around them. We have characterized wonder so that it encapsulates this constellation of cognitive activities and binds them together with a motivational passion to understand and thereby see the world in ever more rewarding and insightful ways.

Preschoolers are eager to learn about topics that adults call parts of science and engineering. Their enthusiasm and joy in learning is obvious.

When our own sons and granddaughters were infants, they were constantly trying to learn more about everything they encountered. Whenever possible, they tried to manipulate objects, or intervene on systems so as to better understand them. But the most salient behavior of all was how much they simply enjoyed the process. A six-month-old cares little about material possessions, status, or power, but after food and social support are supplied, one of their strongest remaining passions may be simply to gain better insight into the world, experiencing, and then creating, aha moments.

All these early abilities, impressive as they are, gain much more power when they are embedded in worlds populated by other agents with their own unique knowledge bases. The next chapter turns to how we amplify our understandings through interactions with others.

3 Working with Others

Many years ago, after observing two caregivers over several months in a day care center that our son Dylan was attending, I started to notice a recurring pattern. Both women appeared to be equally warm and liked by the children, but when problems needed to be addressed, the children went to different adults depending on the problem type. If a new toy stopped working, or if a door was stuck, they would go to one adult, but if the hamster seemed sick, or a plant was dying, they went to the other. They seemed to sense that these two people had complementary areas of expertise and could use those intuitions to guide which one to approach for a novel problem with a new kind of entity (e.g., a hamster) in the same general category (e.g., living things). Of course, these were just casual observations of behaviors that may have occurred for very different reasons, and it was not until almost two decades later that our research group conducted systematic experiments to show that preschoolers do have intuitions about areas of expertise and what those areas mean. Most critically, the ability to extend our understanding through access to the contents and products of other minds enhances wonder. We gain immensely more powerful ways of speculating, exploring, and discovering.

While we may all realize that we sometimes learn from others, we underestimate how much we depend on other minds. We labor under an "individualism bias" in which we fail to acknowledge how deeply we are immersed in, and depend on, knowledge communities.[1] This nearsightedness about others' contributions, which is considerably worse in children, not only masks the full extent to which young children depend on other minds but it also underestimates just how skillfully they can navigate the terrain of knowledge. The psychologist Paul Harris once said that we tend to mistakenly think of children as "stubborn autodidacts," resolutely teaching themselves without assistance.[2] This skill at tapping into what other experts know and learning how to rely on them appropriately is at the heart of successful science. It also opens up new dimensions of wonder as it illuminates new puzzles and new opportunities for exploration and discovery. Several recent studies illustrate how elements of these expert navigation skills emerge early in children even though the same skills can become ignored or neglected by adults.[3]

We have already encountered some indications of children relying on information that they could have only learned from others and not through direct experience. Even in the preschool years, children acquire information about the world from books and other media. By necessity, much of their early folk biology comes through socially transmitted information. It is the only way to learn about germs, internal organs, or dinosaurs. They heavily rely on others to learn how to use various devices such as television remotes or jack-in-the-boxes and often learn important information about invisible parts such as batteries. This happens so often it may seem to be a trivial information-gathering process, but it involves a diverse set of skills that, if not mastered, would make learning from others hopelessly hard.

Consider some things that an infant or young child must do to effectively learn from another person. They must choose an informant. They must decide what information that informant can reliably provide. They must make inferences about the reliability of that information itself. These inferences depend on general properties of informants (e.g., the quality and relevance of their knowledge, their motivations, their areas of expertise) and the nature of the information (whether it is intrinsically unreliable or not, such as long-range weather predictions vs. sunset predictions). Children must understand how the informant acquired the information. How hard would it be to acquire that information on one's own in comparison to asking an informant? They should consider what questions to ask informants and how to productively argue with them. This list is hardly exhaustive, but it illustrates some of what is needed for children to fully benefit from the knowledge communities that surround them. Those benefits in turn enable them to contemplate more fully and clearly what else they might learn about.

Why do all these skills at learning from others matter for wonder? Wonder can be greatly amplified and extended through interactions with other minds. On our own, we may wonder why gold is so resistant to corrosion.

When we find an appropriate expert on metals, we learn that gold is in the class of noble metals that don't easily corrode. We are then enabled to wonder what is shared by the noble metals that enables resistance to corrosion, and, through our expert, we may learn about electron bands and how chemical properties of elements with certain filled bands are different.[4] Each successive interaction with another gives us a sense of whole new avenues of inquiry.

Learning from What Others Do

One form of learning from others is not wonder-driven itself but fuels wonder in a manner analogous to statistical learning by ultimately providing a supply of new puzzles. This can happen when we learn through observing the actions of those around us. Through such observations we pick up cultural routines and practices.[5] We can learn from models everything from specific aggressive behaviors, such as punching a Bobo doll in the face, to walking styles.[6] We don't copy every action we see. We don't generally copy sneezes, stumbles, or obvious failures. Copying seems to be a strategic way to selectively learn skills and how to behave in culturally normative ways. However, when we copy the actions that people perform on other things, we can appear quite incompetent at times.

Humans, unlike all other animals, faithfully reproduce actions of others even when those actions don't make any sense. This phenomenon, which Derek Lyons and I labeled "overimitation," occurs when preschoolers observe someone manipulating parts of a complex device to achieve a goal. In most experiments the goal is retrieving an attractive object from the device's interior.[7] The actor performs some actions that seem obviously necessary and others that appear to be irrelevant and silly. For example, they might (1) move a latch on top of a door that is near the toy and (2) tap a rod twice on a part of the device far from any access point to the toy (see figure 3.1). Preschoolers will imitate both actions while chimps and dogs tend to imitate just the reasonable first one.[8] Figure 3.1 illustrates the causally relevant and irrelevant steps for a puzzle box originally used by the psychologists Victoria Horner and Andrew Whiten. Only when the actions are so silly that they violate the strongest principles of early "core" knowledge, such as manipulating a part of an object that is completely detached from the device, do children resist overimitating.[9]

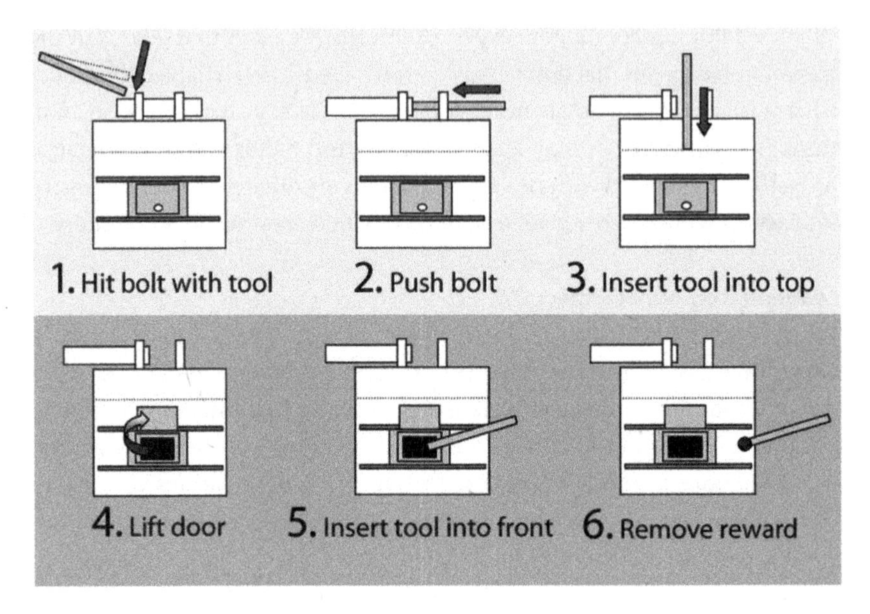

Figure 3.1

Overimitation. Young humans copy both the steps that seem causally irrelevant retrieving the desired object, as shown in the top panel, as well as those steps that seem causally relevant as shown in the bottom panel. In contrast, primates only copy those steps that seem causally relevant. (Adapted from V. Horner & A. Whiten. (2005). Causal knowledge and imitation/emulation switching in chimpanzees (*Pan troglodytes*) and children (*Homo sapiens*). *Animal Cognition, 8,* 164–181.)

Why do chimps and dogs appear to be more sensible than humans? The answer reveals a deep truth about human learning: humans assume that others can reveal interesting, even counterintuitive, properties of devices through their actions. They believe that humans tend to operate efficiently on the world and therefore their actions must be causally meaningful even if at first glance they don't seem to make sense. Remember how you feel when a friend shows you how to use a complex new device such as a weaving loom. As your friend demonstrates, many steps seem to make no sense. Nonetheless, you immediately assume that they are indeed critical and copy them while you try to learn. You don't normally assume that friends will engage in lots of unnecessary actions to obtain a goal.

Humans put great faith in the information value of others' actions. This makes sense given the immensity of human cultural knowledge in comparison to that of all other species. As much as we might appreciate the information passed on by bees through their dances, birds through their songs, or monkeys through their calls, no other species comes remotely close to us in terms of the amount and diversity of information shared. Humans may have developed a special system for learning about initially opaque causal constituents of complex devices. Overimitation starts functioning in humans before age two. It initially occurs rarely and then reaches a plateau at age five or six. It then continues at roughly the same level throughout life.

Beyond supporting learning of causal relations, overimitation can occur for social reasons.[10] Sometimes children overimitate to indicate social affiliation, to adopt a convention, or to learn a ritual. In such cases, they may know that the extra actions are causally silly but choose to imitate them anyway because the social context suggests that doing so carries important social benefits. Consider the various rituals members of some groups engage in when they greet each other, ranging from extended handshake interactions to highly routinized sequences of back-and-forth bantering. This affiliation signaling helps explain why overimitation is more common in response to members of one's in-group than in response to an out-group.

Overimitation is best understood as a uniquely human way of decoding complex actions produced by others in the service of two kinds of goals. One goal involves learning how to act on tools, devices, and other artifacts. It works through noting and remembering all the components of an action sequence with the assumption that each plays a critical causal role in working with the object. This form of overimitation promotes wonder. A second goal involves learning how to fit in with others in a community through rituals that signal and unify. Overimitating helps children decode the critical "syntax" of ritualized social events. In both cases we accept some steps as important without understanding why. Does one goal precede the other in development and possibly also in evolutionary history? Given the challenges of figuring out many social norms,[11] overimitation might first appear with tools for which critical but nonobvious manipulations must be learned. In fly-rod casting, for example, an expert caster does a great many small things that make little sense to a novice observer. Failure to copy those actions, however, guarantees a frustrating day standing in a cold river. Overimitation helps us appreciate the causal relations inherent in a device that are not apparent from casual inspection. We gain insight by taking a leap of cognitive faith, namely that intentional actions on objects are likely to reveal important things about those objects even though we don't yet understand why. Over time, we start to ask why we do superficially silly steps. What is it about the device at a deeper level such that tweaking a seemingly irrelevant knob is really essential? As these questions emerge from introspecting about our actions on any object, they set us up to wonder about the object in new ways. What hidden causal factors can explain why that odd action is actually critical to understanding how that device works?

Learning from What Others Say

We often learn from others when they are providing explicit demonstrations with lots of accompanying language. In these pedagogical situations, young preschoolers make several inferences. For example, they might be taught how to use a device with several manipulatable parts by watching a teacher manipulate a few of those parts to achieve a desired result. When they later are left with the object on their own, in contrast to untaught children, they show less spontaneous exploration and discovery of other properties that weren't touched by the teacher. As a result, they don't learn the consequences of manipulating those other parts.[12] Imagine that a child's parents just brought home an expensive new robotic vacuum and they show their preschooler how to open a small door and flip a switch that turns on flashing warning lights. That child is considerably less likely to learn about the vacuum's more important features than his sister who later stumbles upon it alone and starts exploring its buttons and levers. Thus, when teaching is too heavy-handed or doesn't provide for open-ended learning about additional features, it can stifle rather than inspire wonder.

When to Ask Questions

Children start asking questions early on in language acquisition, and the rate of questioning rapidly grows, as does the variety of question types. The earliest questions before age two are often information seeking. Children ask about what things are called, where they are, and what properties they have (is it hot?). They can ask more than 100 questions per hour during the preschool years. The first explanation-seeking questions usually appear well before age three but are rare in the beginning. Explanation-seeking questions such as why and how questions then emerge in full force around thirty months, after which time roughly a third of preschoolers' questions ask for some sort an explanation rather than for a mere fact or to make a request.[13]

What do children really want to know with their why and how questions? Some clues can be found in the adult motivations behind what the psychologist Tania Lombrozo and colleagues have called *explanation seeking curiosity*, or ESC.[14] ESC is the drive to obtain explanations and not just facts or directions on how to do something. While questions in general usually involve other people, you can also seek explanations from yourself, which turns out to be a surprisingly effective learning strategy.[15] Which adult motivations for seeking out explanations are also found in young children? Those motivations include surprise, novelty, ambiguity, and information gaps. If we sense a hole in our understanding, we have a need to fill that hole. A final adult motivation is the expectation of valuable information later on. This motivation focuses on future payoffs of gaining explanatory insight and not on benefits already gained. For example, if you decide to learn about the human microbiome, the bacteria, viruses, and other microorganisms that normally live inside us, you may do so based on vague hunches about future benefits for understanding human health and disease.

The ability to recognize gaps in knowledge and future benefits of understanding may gradually develop during the preschool years. Some forms of ESC may require an awareness of one's ignorance as well as a sense of how to remedy that ignorance and why doing so might provide future benefits. These introspective skills continue to be refined well into adult life. But, in highly familiar areas of knowledge, children as young as two may know that they lack a particular part of understanding about how a toy works and that if they did know it, they'd be able to do more things with the toy.

In short, question asking rises dramatically between ages two and three both in terms of raw frequency and in terms of asking more questions about how things work and why things have certain properties or behaviors. You'd think that all this provides a platform for an even greater burst of questions once children enter real classrooms. Surely formal education will provide new students with ideas and riddles that will cause them to ask about many more things in much more depth. Yet, that rarely happens. Instead, questions tend to dramatically decline during the elementary school years.[16] Most children continue this unquestioning passivity for the rest of their lives—a deeply disappointing outcome given preschoolers' rapidly improving questioning skills and how those skills propel wonder.

The disconnect between school and home can even happen in preschool. In a set of detailed observations almost forty years ago, the psychologists Barbara Tizard and Martin Hughes found that children in British preschools asked far fewer questions in preschool than in the home. The difference seemed to arise because teachers quickly assumed the role of question asking and assigned to children the role of question answering.[17] At home, young children initially explore everything around them through their actions, a kind of nonverbal wondering. After two years, why and how questions start to emerge in concert with those actions, but soon, questions come in a torrent, sometimes at rates of more than one a minute over a two-hour interval. Most of these questions reflect sincere desires to learn more about how the workings of the world and can be less directly linked to concurrent actions. Tizard and Hughes describe these children as "puzzlers" eager to extend their understanding through conversations with their parents. They called many of these conversations "passages of intellectual search," in which children build an understanding over time through a sequence of interactions with their parents. Teachers in even the best preschools rarely have the time to construct such conversation-driven passages for each child. Instead, they come to monopolize the question asking, a practice that becomes even more dominant in most elementary schools.

To question effectively, you should know when to ask questions and whom to ask. We ask when we recognize gaps in our knowledge and believe someone might close those gaps. Wonder inspires the most productive and fruitful lines of questioning by going beyond simple requests for facts or for clarifications. Moreover, asking questions is effective. It can increase the elaboration of an explanation under consideration by a child and an adult.[18]

How do children recognize a gap and the need to ask others? Can they inspect their knowledge and see where it falls short? Illusions of knowing can cause even adults to stumble here. Introspecting about your cognitive states is known as *metacognition*, namely thought about thought or stepping back from the use of knowledge to inspect that knowledge. This is an especially difficult task for preschoolers.[19] Metacognitive performance depends on *executive functions*, loosely defined as processes involved in cognitive control and monitoring. Decades of research demonstrate weak metacognitive skills early on. Preschoolers have trouble judging their ability to remember things they have recently experienced. Similarly, they can be terribly mis-calibrated in predicting their performance on a memory task. They struggle with knowing how and when they originally learned something and with identifying cognitively challenging tasks.

In one study on memory for sequences of cards, four-year-olds guessed how many cards depicting everyday things (umbrellas, dogs, cars . . .) they would be able to remember. They were shown successively larger rows of cards and told to guess, for each larger row, whether they could still remember all the things shown when the row was covered. Older children, and adults, both started to doubt they would remember everything when the number exceeded approximately five cards. Four-year-olds confidently predicted perfect recall when the string was as large as ten cards. When those same children were tested on recall, they remembered only two or three pictures.[20] They were wildly overoptimistic about their memory skills.

Another study taught four- and five-year-olds novel facts about animals, such as that tiger stripes provide "camouflage." Most of the children did not know any of the facts, yet when later asked how long they had known the just-learned information, they often claimed they had always known it or had known for as long as they could remember.[21] Adults can make similar mistakes through *hindsight bias*, but children err much more frequently. However, even when children fail, they reveal sophisticated assumptions about the nature of knowledge itself. They commit the "knew it all along" effect more often for statements about categories of things than for statements about individuals (Dogs get sick after eating carbamates vs. Last night, this dog got sick after eating carbamates). This category effect reflects an early belief that category knowledge is more likely to be common knowledge.[22] If you think a certain kind of knowledge is more likely to be widely shared, you tend to assume you have always known it as well.[23] If you don't know it, you are more likely to ask others if you also think it is common knowledge. Statements about individuals don't carry that implication. These subtle distinctions not only help young children learn from others but they also support wonder. Wondering about categories and kinds is more rewarding and mind-expanding than wondering about individual cases. Similarly, they grasp the increased value of abstracting over a group rather than focusing on the particulars associated with any one individual.

Scaffolding Neglect

Scaffolding occurs when learning is tacitly supported. For example, when teaching a child how to make a bed, you might lay out the bedding materials in a spatial sequence suggesting an order of actions. Perhaps the mattress pad is laid out with one part already around a corner of the mattress. The bottom sheet is laid out next to the mattress pad in a manner implying a similar wrapping around the mattress. Next, you might place the top sheet, followed by a blanket, pillows and pillowcases, and finally by a bedspread. The spatial layout of the items, and their orientations, suggest the order in which they are to be used. A child might watch you lay out the items and listen to you make vague remarks such as "Oh, now we need this." Suppose children easily make the bed with this scaffolding in place. Will they know how they relied on the arrangement of materials? Will they succeed when presented with all the materials mixed up in a laundry basket? Scaffolding implies that support enabling early learning and performance needs to be removed carefully and bit by bit before the learner can function independently, much as a scaffold is gradually taken away from a building as it nears completion. All of us underestimate how much we rely on such supports.

Are young children especially blind to scaffolds? We explored this issue with young schoolchildren who were subtly scaffolded on how to use a novel tool to carve their initials in crayons.[24] With scaffolding, almost every child succeeded in learning to use the carver while none succeeded without any scaffolding. Yet, most children thought that they could have easily learned how to carve without it. Only when children observed another child receive the scaffolding did they realize its importance. When children are totally immersed in learning about how to operate devices, they largely ignore the learning support provided by others. Despite this oversight, these illusions may be helpful to young children. A neglect of scaffolding may increase feelings of efficacy even if those feelings exaggerate what children actually accomplish. Moreover, because explicit pedagogy can dampen spontaneous discovery in children, a blithe ignorance of how much others were "teaching" them may ironically promote more wonder-based exploration.

Support from Nonverbal Metacognition

In contrast to explicit verbal assessments of what one knows, when tests of self-knowledge are nonverbal and less explicit, preschoolers can better sense what they know. This is especially true when the task is embedded in a social environment where other people are potential sources of information.

One experimental paradigm was inspired by studies in which monkeys had to remember food locations. They could choose whether or not to have their memories tested. The monkeys chose not to be tested in just those situations where they had good reason to doubt their memories.[25] The analogous human study combined the hiding manipulation with the tendency of toddlers to point.[26] Toddlers learned to point where they thought a toy was located after watching it being hidden in one of two boxes. When they pointed to a box, it was then pushed toward them. If they were correct, they immediately got the toy. If they were wrong, they got nothing and had to wait for another trial. However, they also had the option of asking their caregiver to tell them where the toy was, a slower but more reliable method. Memory difficulty was manipulated by varying how long after the hiding event the toddlers had to wait before being allowed to point or ask the caregiver for help. The longer the delay, the more likely they were to ask for help than to point directly. They sensed the increasing fragility of their memories over time and knew when it was better to rely on others. In other parts of the study, toddlers who did have longer delays were indeed much worse in memory performance.

Striking disconnects occur when children make explicit confidence judgments instead of revealing their confidence implicitly through their actions. When children are asked how well they can identify drawings of recently seen common things that are depicted in degraded forms (e.g., missing lines), they are highly inaccurate. In contrast, implicit measures, asking for a cue to help them, reveal greater insight.[27] Children as young as two-and-a-half request cues more often when their actual performance is likely to be worse, but not until age four do children explicitly rate their confidence as lower in such cases. They "know" when they are likely to need help because they are accurately assessing their performance ability, yet they are unable talk about that insight explicitly. Metacognitive insights seem to first occur automatically and implicitly with more explicit metacognitions in the same tasks emerging later.

Children in these studies showed increased metacognitive ability when that ability was used to infer when to seek help from others. When embedded in a social situation with others as salient potential sources of information (unlike scaffolding), preschoolers sense their cognitive limitations even as they have trouble with explicit introspection. Knowing when to ask questions may be an early developing skill in contexts where others are demonstrably available to answer those questions and the child does not have to step back and explicitly evaluate their knowledge and ability, but rather merely needs to decide whether or not to ask for help. Such social contexts may similarly promote new ways of wondering. Thus, if you are a young homicide detective and think you have a good grasp of the various ways of gathering evidence, you may suddenly be aware of gaping holes in your knowledge when your senior partner has a forensic entomologist and a forensic architect join your team on a particularly tricky investigation. Neither specialist has told you anything yet, but you may not only suddenly realize huge gaps in your investigative repertoire, you may also start wondering about entirely new lines of inquiry and insight now available to you by having these people in your group and eager to help.

What Is Plausibly Learnable?

A different way of sensing the need to seek help from others involves considering how bits of knowledge are plausibly learned. Across several lines of work, our lab and others have uncovered early abilities to infer what is plausibly learnable. In one line of work, our lab group asked children to imagine a young boy growing up on a deserted island with all the food and shelter needed to thrive but no other people around.[28] They were then asked what that child could learn on his own and what he could not learn because he grew up alone on the island. They were presented with "direct experience" information that could be easily learned by the lone islander, such as "the sky is blue" or "one sleeps when one is tired." They also heard "indirectly acquired" information that the lone islander could not possibly have acquired on his own, such as "germs make you sick" or "stars are hot." Even the youngest five-year-olds easily recognized that the lone islander knew the directly experienced information but not the indirectly acquired information. It didn't matter whether the indirect information was highly familiar or not.

Finally, children as young as five could tell that some information could easily be acquired firsthand, such as "Some animals are awake during the day but sleep at night, while other animals are awake in the night but sleep during the day." They further sense that some other kinds of direct knowledge could be acquired firsthand, but only with great difficulty, such as "Ants walk in a zigzag when searching for food, but walk in a straight line when going back home." All children judged the lone islander as less likely to acquire the difficult knowledge. At the same time, they saw the lone islander as more likely to acquire this difficult knowledge than necessarily indirect knowledge about such things as dinosaurs. Knowledge such as that germs make you sick and that stars are hot could only be acquired from others.

Given the metacognitive literature about explicit judgments, we were surprised at how easily all ages distinguished the relative difficulties of acquiring the different kinds of knowledge. As in the scaffolding case, reasoning about others may be easier than introspecting about one's own knowledge. Developmental change did occur in one respect. While the youngest children definitely thought that the indirect knowledge items were less likely to be acquired by the lone islander than the direct items, they often believed that some indirect things might be possible with great difficulty. They seemed reluctant to acknowledge the absolute impossibility of learning about germs on one's own. However, in other tasks in which they directly judge the possibility of events, young children tend to see the unlikely as impossible.[29] Thus, early judgments of the possibility of gaining some kinds of knowledge shift the impossible to the unlikely. In contrast, judgments of the possibility of events themselves shift the unlikely to the impossible.

Young children also have a sense of when something is too complex for them to learn on their own. Our research group has explored how children's intuitions about the complexity of objects are linked to their sense of needing help to plausibly learn about those objects.[30] By age seven, children's judgments were largely in synch with adult intuitions about the mechanistic complexity of things ranging from the very simple (hammer) to the very complex (jet airplane). They used those intuitions to guide judgments about whether they would likely need the help of an expert to understand how the device worked. These widely consensual intuitions were especially impressive given that the children had little to no knowledge of the mechanistic details of any item, just a general impression of the causal complexity underlying the devices. Complexity intuitions suggest new avenues of wonder as well. A sense that a class of things operates by means of complex internal causal processes that some group of experts understands can easily lead to conjectures about what those processes might be and how to pose the best questions to those experts.

Younger children are not always well calibrated. They do envision acquiring exceptionally high levels of knowledge. For example, when anticipating how much they or a peer could learn by the time they became adults, younger children thought they would have extremely comprehensive knowledge. Older children and adults were more conservative in judging how much they would know.[31] The younger group saw their own knowledge as rising faster than for peers and saw moral knowledge (e.g., knowing that something was wrong) as especially easy to acquire. This optimism about future knowledge reflects a broader pattern of youthful overoptimism. Kristi Lockhart proposed that early optimism may have adaptive value for children because it reduces discouragement after experiencing early transient failures that will soon be outgrown.[32] This may let them speculate about the unknown less tentatively than older children and adults.

Whom to Learn From

A large literature on "trust and testimony,"[33] much of it arising from the psychologist Paul Harris's extended research group, describes preschoolers as carefully evaluating what others say. They rely on statements made by informants depending on informants' past error rates, their confidence, their relevant experience, and even their overall warmth. An especially powerful and enduring cue may be consensus of others with the informant.[34] Using these strategies, preschoolers rely on those informants that seem the most credible. Skill in selecting sources allows you to more adeptly integrate them into your own thoughts about how the world might work. If you are able to easily access people who both know relevant content and who sincerely want to assist you, your ability to appreciate what lies beneath first impressions will soar.

Children don't blindly absorb anything they hear, even though adults often think they are hopelessly gullible. The concept of omniscience develops surprisingly late. Even when told that a person "knows everything about everything," children under seven put boundaries on that person's knowledge. Preschoolers judge that experts have much greater depth of knowledge than the supposedly all-knowing person.[35] The notion of experts having a privileged knowledge in their own areas is so natural and compelling that it falsifies the logical truth of omniscience. It can even overrule simple truth telling. We have found that children as young as four can tell that true but irrelevant responses to questions are less helpful than true and relevant answers.[36] In certain cases, they knew that, just because someone says something that is true, truth alone is not enough to qualify that person as an expert.

While they are able to selectively trust sources, younger children are not as cautious about misinformation as others. The psychologist Candice Mills found a marked vigilance shift between ages five and eight. Younger children assume that someone who claims they won a race is just as believable when winning guarantees a big prize as when it does not. In contrast, older children and adults suspect that claimants may deliberately, or perhaps even unwittingly, be biased toward making statements that benefit themselves.[37] Younger children tend to be dewy-eyed idealists about others' statements. They are not natural born cynics. They find it similarly difficult to see partiality in judges. They don't realize how motivations might cause judges to distort assessments of the actions of others. Those children are certainly aware of negative and positive attitudes, but they often don't see such attitudes as grounds for distrust.[38] Motives are challenging to integrate with assessments of truth.

A related process occurs in developmental shifts in views of boasting. Kristi Lockhart led a project showing that younger children construe boasting more benignly than older children and adults. Children aged seven and younger do not regard boasters negatively. Instead, they see braggarts in a positive light and as socially sharing useful information about their skills. This is a striking example of younger children's inability to combine ulterior motives with the content of the message.[39] Adopting such a benign attitude may not be as risky as it seems when children are young. They usually spend their early years surrounded by parents and other well-meaning caregivers. Only later do they encounter strangers and peers with more malevolent intentions. This growing suspiciousness and distrust may come to suppress wonder as children grow older.

We frequently evaluate our need for informants by consulting our intuitions about the topic at hand. Preschoolers can combine their own sense of patterns in the world with the quality of a person's statement.[40] They can consider physical plausibility and possibility when evaluating whether a description is true. For example, in deciding whether the protagonist in a story is a fantasy character or an actual character, they consider the physical plausibility of their actions.[41] If a teacher demonstrates how a toy works but focuses on just one of several critical features that a six-year-old knows about, the teacher is considered less trustworthy.[42]

The degree to which children doubt a claim because it contradicts what they have learned about the natural world is influenced by culture as well. When children grow up in deeply religious cultures (ranging from Christian to Islamic), they are more willing to accept claims that violate their intuitive beliefs, such as miracles.[43] They seem to regard fundamental causal patterns in the physical world (including biology) as strong and reliable but still capable of being overridden by appropriate supernatural agents. This tolerance of supernatural phenomena can continue into adulthood, requiring the curious coexistence of both natural and supernatural beliefs.

Evaluating Explanations

In addition to evaluating testimony in terms of its compatibility with regularities in the world, children soon learn to evaluate statements in terms of their intrinsic value and informativeness for the task at hand.[44] An important class of such statements is explanations. Someone who provides a poor-quality explanation will drop in favor as a trusted source on the specific topic area and, often, on other topics as well. Preschoolers and young schoolchildren are aware of several factors related to the quality of an explanation: circularity; simplicity, and generality.

Even kindergartners realize that someone who gives an empty circular explanation doesn't know as much as someone who gave a noncircular explanation of the same thing.[45] In one series of studies, we described scenarios of the following sort to children: "These people think they know how a can opener works. Which one do you think knows the most?" One person is described as saying, "A can opener works because it does something that takes the lid off so that when you are finished using the can opener the top of the can has come off" (circular). The other person is described as saying, "A can opener works because it has a sharp wheel that cuts the lid and another wheel that goes under the lid of the can and turns it so that it can get cut" (noncircular). Kindergartners choose the noncircular statements for several different devices and animals. However, their ability was fragile compared with that of older children. They had difficulty with circular cases if they didn't also see a shorter circular statement on the same topic ("A can opener works because it opens the can"). The shorter examples were more blatantly useless and helped younger children see the problems with longer circular statements. Yet, when they were asked to justify their judgments, they didn't refer to the answer length or complexity but instead focused on how the noncircular case revealed more useful information.[46] Later studies suggest sensitivity to the weaknesses of circular explanations in three-year-olds, and even, in one case, two-year-olds.[47] In many cases, embedding tasks in social learning contexts results in demonstrations of more precocious metacognitive abilities.

Children also seem to prefer simpler explanations. For example, children at least as young as four tend to explain multiple outcome events in terms of a single common cause and not in terms of two independent causes. This bias holds even when they understand that the multiple cause scenarios are just as likely.[48] Adults sometimes prefer more complex explanations if they think such explanations will fit the observed pattern better,[49] but children may have a cruder overall simplicity preference.

An appreciation for more general explanations seems to appear at around age five. Assuming that more general explanations have more power and offer greater insight, one study pitted "generic speakers" against "specific speakers." The generic speakers made statements about a category of novel animals ("Pangolins sleep in hollow trees"). The specific speakers made statements about an individual animal ("This pangolin sleeps in a hollow tree").[50] By age seven, children favored generic speakers as being more knowledgeable not only about the taught animal but also about other animals and their properties. Five-year-olds showed signs of a generic bias but only when the statement was perceptually obvious, as in "Pangolins have pointy noses." Thus, a preliminary sense of the value of general statements emerges before the start of formal schooling, but it becomes more robust over the next two years. All school-age children also see people making generic statements in support of explanations as more knowledgeable and better sources than those who make specific statements.

More broadly, children appreciate the virtues of generality in explanations, much as they do with questions. We found that preferences for generality emerge by five years for biological phenomena (e.g., "All animals have muscles that stretch and shrink to change the position of their bodies") but not until around age seven for physical phenomena (e.g., "All things leave a shadow when they keep light from passing through them").[51] This difference may arise because children have learned richer general facts about biological kinds, such as kinds of animals and kinds of plants, than they have about different kinds of physical things. Other assumptions may further guide young children's intuitions. Thus, statements that apply to more instances will be more valuable, but they can't be excessively vague. If you say that "Pangolins are things," that statement is no more informative than if you say "This pangolin is a thing." The content has to provide useful information.

In short, most young schoolchildren can evaluate sources in terms of the intrinsic qualities of the explanations and statements that they provide. Combined with the ability to evaluate informants in terms of how well their statements fit with known real-world patterns, young children have a diverse array of tools to guide them in making judgments about whom to trust. While overestimating how much they can learn on their own, they still recognize the need to selectively rely on others. With development, they become more discerning judges of informants and how their motivations should be taken into account,[52] but even infants sense that some people (e.g., parents, or those with a similar accent) might be more reliable learning partners than others.[53] This early ability may help them sense those situations in which wondering will be most welcome and most leveraged by the presence of interested others. All of this sets up one of the most profound products of human activity: culture and the divisions of cognitive labor that comprise all cultures.

Children cannot fully benefit from what others know if they do not have some sense of how expertise is distributed in the minds around them. Even if they have never heard of a discipline such as chemistry, they would benefit enormously if they were able to identify those who were likely to be more expert in that area. Are there ways they might be able to identify such experts before they even set foot into a classroom? To better understand how this might occur and how such abilities might tacitly support us on a lifelong basis, we step back for a moment and consider why and how labor is distributed in a society.

The Division of Cognitive Labor

The origin of classical economics is often attributed to the Scottish economist and moral philosopher Adam Smith in his magnificent 1776 book "An Inquiry into the Nature and Causes of the Wealth of Nations."[54] Smith argued that the wealth of nations is closely related to increasing divisions of labor and that those divisions enable economic growth. Thus, as members of a society specialize, they achieve greater mastery in their areas of specialization. As a result, they become more proficient and add more value to their labors by producing better quality goods and services.

This evolving specialization is easily experienced in contemporary society when one moves from a less affluent region to a more affluent one. When my family moved from a small rural town near Ithaca, New York, to a larger Connecticut town near New Haven, we were stunned at the increased specialization levels of those doing home renovations. In our Upstate New York town, it was often the case that a single person served as the "contractor" who also performed aspects of the job from pouring the foundation to painting, with occasional support from helpers such as a backhoe driver or a sheetrock assistant. When we wanted to do a modest renovation in Connecticut, the contractor never did any direct work on the project and instead hired over thirty different specialists. For installing windows, one group framed the window openings, a second group put the windows in the openings, a third hung the sheetrock around the windows, a fourth taped the sheetrock, a fifth painted the sheetrock, a sixth installed the trim around the windows and a seventh stained and varnished the trim. Seven specialists had replaced one.

An implicit assumption behind Smith's account is that divisions of physical labor almost always correspond to divisions of *cognitive* labor. The philosopher Philip Kitcher first used the phrase *division of cognitive labor* in a 1990 paper on how mental labor is apportioned in the sciences.[55] Fifteen years earlier another philosopher, Hilary Putnam, discussed the *division of linguistic labor* and how the meanings of many terms depended on experts.[56] For most people, the meanings of words such as *water*, *gold*, and *tiger* depend on the presence of experts who can tell us when *gold* correctly refers to the real metal as opposed to imposters such as fool's gold. Advanced cultures are comprised of vast numbers of different communities of relative experts who exist in highly interconnected webs.

These divisions of physical and cognitive labor are not restricted to recent high-tech cultures; instead they are intrinsic to any sense of culture at all. Even the most traditional human societies, such as hunter-gatherers or food producers, have individuals who specialize more in one skill than another. This often includes divisions of labor revolving around gender, but, within a gender as well, some are more expert in healing, others in hunting, and still others in foraging. Those distinct skills suggest different forms of cognitive expertise as well. In the vast majority of modern societies, the divisions are much more elaborate and run much deeper. The general ophthalmologist of forty years ago is now much less common in affluent communities, having been replaced by specialists in areas such as retinal disorders, cataracts, corneal injury, tear duct disease and LASIK.

If divisions of cognitive labor have been important parts of social groups for several millennia, then the ability to navigate such communities of knowledge might be an especially valuable cognitive tool and one that opens up vast new territories of wonder. We constantly defer to others who have greater expertise whether they be health practitioners, auto mechanics, or river guides. We defer many times a day, yet this ability has only recently been an area of focused research. It is also at the center of modern science. Virtually all scientists are deeply dependent on others' expertise to be able to carry out their own research. The biologist who studies cell movement must know whom to rely on when using electron microscopes, modeling the physical strengths of membranes, or uncovering gene regulation pathways. These expertise navigation skills vital to practicing scientists are foreshadowed by precursor abilities present in preschoolers and young schoolchildren, and these early abilities often function outside of awareness. Moreover, such skills are rarely targets of instruction at any age. Without any instruction, university students can judge whether an expert's area is pertinent to a science topic and apparently succeed by having acquired a rough implicit sense of disciplines.[57] The tacit nature of this skill in adults suggests much earlier origins. What elements of this ability are present before schooling starts?

Young children not only have a sense that different groups of other people have greater levels of distinct mastery, they also use quite subtle and sophisticated strategies and heuristics to support those impressions. In one set of studies we told preschoolers that "Anne knows all about eagles such as the kinds of food eagles eat, how many babies they have, and how big they can grow," and that "Betty knows all about bicycles such as what they are made out of, how to fix them and how their brakes work." We then asked them who knows more about other topics. Children as young as three infer that Anne also knows more about not just other birds such as ducks but also all animals and perhaps even plants. Betty in turn knows more about not just tricycles and cars but even about door locks.[58] These are examples of the sorts of experimental materials that confirmed what seemed to be going on in the day care center described at the beginning of this chapter.

We then explored the growth of knowledge navigation skills through the elementary school years as well as in adults.[59] Even kindergartners saw much of expertise as organized roughly like the natural and social science departments of a modern university. For example, someone who understands why big boats take a long time to stop was judged by the children as more likely to know why basketballs bounce better on the sidewalk than on the grass (physics) than why ice cream cones cost more in the summer than in the winter (economics). Since these children obviously have never heard of those departments or those fields, how are they succeeding? We found that even the youngest children were not simply looking at familiar objects and finding surface similarities to other objects or situations. Instead, they assumed that those who had mastery in a domain grasped some deeper causal pattern (e.g., about how inanimate objects move and interact with each other). This insight then enabled them to see how other phenomena that are different on the surface, such as those involving boats and basketballs, arise from the same underlying causal regularities (see figure 3.2).

As they progress through the early school years, children refine these skills and focus on deeper and more elaborated causal patterns. They may also tend to shift in their intuitive ideas of expertise areas from goal-oriented categories (e.g., fishing) to more discipline-centered categories (e.g., biology).[60] They first focus on expertise organized around practical skills that address important challenges, such as finding food or healing an illness. Over time, however, they come to see that several diverse goal-centered practices may be relying on a shared set of causal patterns common to a discipline, such as those that govern biology. They also learn what it means to have "expertise" available on the internet through search engines. Today's children frequently watch their parents put questions to a speaker sitting on a bookshelf and hear it quickly respond with answers. Such disembodied sources of information require a whole new set of skills to understand how to evaluate the knowledge they deliver.[61]

From preschool on, children know that the partitioning of knowledge is linked to different kinds of causal patterns in the world. Children sense that when someone grasps those deeper causal patterns underlying one

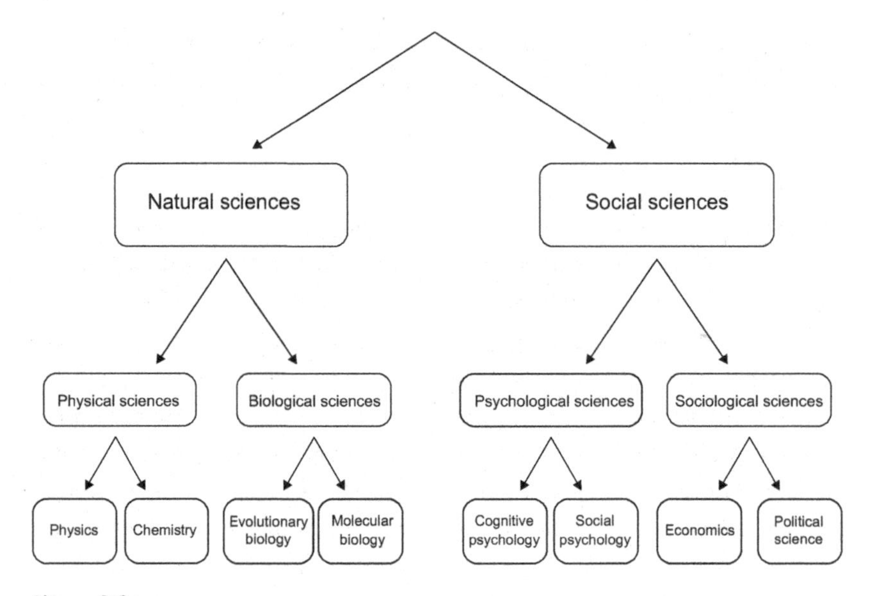

Figure 3.2
Grasping the division of cognitive labor. Despite having never heard of the typical departments and academic divisions of the modern university, children as young as five are able to infer relative knowledge overlaps in the natural and social sciences, as represented by links traversed between any two fields in this hierarchy. Thus, a person who knows all about one phenomenon in physics (limited here to mechanics) is seen as more likely to know more about another physics phenomenon than someone from another discipline even when the two physics events seem very different at the surface level. In addition, experts in physics were judged to have a better chance of knowing something about a phenomenon in chemistry than one in economics. Divisions in the humanities do not fit into the same hierarchy as they tend not to be organized according to underlying causal patterns. (Adapted from F. C. Keil, C., Stein, L. Webb, V. D. Billings, & L. Rozenblit. (2008). Discerning the division of cognitive labor: An emerging understanding of how knowledge is clustered in other minds. *Cognitive Science, 32,* 259–300.)

phenomenon or thing, that same person will have greater insights into an indefinitely large number of other entities and situations governed by the same causal patterns. This is arguably how the modern natural and social science departments of the university evolved. Humanities departments may have arisen in different ways as they are less organized around underlying causal patterns that explain the surface properties of topics of expertise. As study after study uncovered sophisticated and untaught senses of the division of cognitive labor in the natural sciences, we also found a privileging of mechanistic explanations, namely those that break a system down into components that are linked together through a series of reliable causal patterns. Indeed, mechanistic mindsets may be the key to understanding and engaging with science.

The range and depth of wonder expands enormously when we can navigate the division of cognitive labor, even if those navigational skills are rudimentary in the early years. The division of cognitive labor provides far more than the boost that comes with knowing when to trust testimony. It involves an appreciation of the interrelated *structure* of knowledge communities. It reveals hierarchies of knowledge clusters and how they are supported by distinct sets of causal patterns. As a result, we can think ever more deeply about different kinds of possibilities when trying to figure out what is responsible for a phenomenon. To better sense what an eye-opening experience this can be when encountered for the first time, imagine a new student arriving at a university for the first time. Assume the student grew up in a remote region and was homeschooled by his parents. His parents taught him how to read, but their entire small collection of books, which were left by a tourist many years earlier, were regional birdwatching guides. His skill at reading and writing, even if it was just about birds, resulted in his admission to a great university. Now, he arrives on campus knowing nothing about the university except that it is famous as a great place to learn, and he assumes that all learning will be about birds.

On his first day, he is given a course guide and a map of the campus and encouraged to explore. As he wanders about and asks a few questions, he rapidly learns how the divisions of cognitive labor translate into departments and even into areas within those departments. He eventually finds a group that studies birds, focusing primarily on their evolutionary origins. But he then discovers groups in several other departments who study the physics of bird flight, the social interactions of birds, the physiology of birds, the health threats some birds pose to humans, and the use of birds as chemical detectors. World after world of bird investigations open up before him, and he finds himself asking new questions inspired by the existence of those areas. He even starts to think about how other departments might have bird experts as well. He finds himself in a constant state of wonder that soon spills beyond birds to other topics across the university. I had the distinct pleasure of knowing such a student years ago, not quite as isolated from all other areas of study, but close. To watch him revel in what he could explore and discover was one of the most rewarding experiences I've had within academia. Now consider the possibility that all of us as young children had similar revelatory experiences when we began to perceive the divisions of cognitive labor arounds us.

A final aspect of wonder that expands in groups is its transactive nature. When wonder occurs in groups, each member amplifies each other's thoughts through their interactions. I might wonder about how people can steer parachutes. A member of the group might speculate that it is through releasing air on the inside of the chute. I might respond that they'd lose altitude too fast. As various members of the group go back and forth in a cooperative fashion, they consider a much larger range of possibilities and their likely limitations. In a cooperative group with a shared desire to get to the best explanation, this kind of dynamic interpersonal collective wonder, "transactive wonder," can result in extraordinary discoveries and advances. But for many of us, it may happen all too rarely, quite possibly for reasons related to how we neglect positive forms of arguing.

When Others Differ and How to Argue with Them

Arguing is often viewed as interpersonal warfare with the goal of destroying and humiliating the other, or at the very least, the goal of winning. Catalogues list many books on how to win arguments. A few book titles are illustrative: *How to Argue and Win Every Time: At Home, at Work, in Court, Everywhere, Everyday*; *The NEW Art of Being Right: 38 Ways to Win an Argument in Today's World*; and *More Secret Tricks to Win Any Argument.* Somewhat more alarming titles include *How to Win an Argument (Even When You're Wrong)*; and *How to Argue with a Liberal and Win!* In contrast, there isn't a single book title in current catalogues even vaguely similar to the idea of arguing to learn.[62]

Arguing to win has been described as the primal developmental state, one that adults fall back on when under stress or time constraints.[63] While it is true that young children view many disputes between others as attempts by each side to win, the very framing of arguments as disputes obscures a more productive and cooperative side of engaging in arguments. Arguing can also promote learning. Indeed, in the formal sciences, researchers are expected to argue for a specific view or a prediction yet also be fully receptive to a counterargument and indeed use that process as a way of gaining further understanding. Arguing can be deployed as a tool for persuasion and confirmation of preexisting beliefs, but it can also be a mutually beneficial tool for learning and discovery.

Arguing has acquired an undeservedly negative reputation. Alarming numbers of college students are afraid to "argue" for fear that they will alienate others and hurt their feelings. They also worry about the resulting psychological trauma. This anxiety has led to "safe spaces" on campuses where threatening arguments are prohibited.[64] Such attitudes are unfortunate when they create negative attitudes toward arguing in general. Arguing can be an enjoyable and productive way to learn and refine our understandings, as many children spontaneously demonstrate.

At the highest level, skill in argumentation involves using strategies that support the essential features of a good argument. The psychologist Deanna Kuhn has inspired a large body of research showing that middle schoolers, as well as many adults, fail to effectively deploy optimal argument procedures.[65] For example, when middle schoolers are put into competitive situations where the goal is to argue for their own view, they often support their own position instead of refuting their opponent. They also neglect supplying evidence to undermine their opponent's position. They recognize the value of those practices when they are pointed out but fail to spontaneously use them. These strategies have to be used many times in quick succession as an argument between two people proceeds. This process puts large cognitive loads on both parties as they try to keep track of what has unfolded and how to use that information most effectively going forth.

In Kuhn's research area, arguing is typically a competitive process of trying to have your own point of view win. It depends critically on considering all the details of an opponent's position and attacking both the reasoning and the evidence. This process applies to explanations as well. You can be convinced by an intuitively appealing but flawed explanation if you fail to consider supporting evidence. Biases that guide explanation choice could mislead children and adults if they are not also supported by evidence.[66]

At a less fully strategic and systematic level, children do track arguments and evaluate their quality. Two-year-olds favor those adults whose stated reasons for an object's name are better justified. They are evaluating the quality of "arguments" made by others. Moreover, young children certainly can disagree with others and give reasons for their position and even against the alternative. When children choose to not comply with a parent's request, by age five, they effortlessly provide supporting reasons for noncompliance and for problems with compliance. Children as young as two will use reasons to support their point of view in conflicts with siblings.[67]

Science progresses through collaborative argumentation that seeks the truth rather than victory. Consider how a successful research group works together over time. The psychologist Kevin Dunbar and his team spent over a decade observing groups in several biology labs. They studied how lab members interacted in the course of making significant discoveries. Reasoning in these lab groups often led to different individuals drawing contrasting conclusions. These disagreements served to clarify different dimensions of a problem and thereby revealed more accurate representations of the research problem and more plausible solutions.[68] Group composition influenced whether or not reasoning together conferred any advantages over reasoning as lone individuals. If all members came from a very similar background and had worked on the same topic with the same organism, they might fail to come up with critical insights when thinking about a novel problem with a different organism. But groups conferred no advantages over individuals when diversity resulted in different and competing goals. Such situations often evoked "argue-to-win" behaviors as well as blame and responsibility shifting. Dunbar saw this kind of divisiveness as part of the decision-making process that contributed to the space shuttle Challenger disaster.

While an emphasis on arguing to win often overshadows arguing to learn, across many research traditions and disciplines, both kinds of argumentation are ultimately recognized.[69] If both types of arguments do occur in adults, several questions arise: Are there cognitive consequences of adopting one argument type over another? If so, are those consequences relevant to learning about science and preserving wonder? Can we bias people toward one kind of argument without explicitly telling them to argue that way? Is one form of argument developmentally privileged? Does arguing to learn emerge late and for an especially mature "enlightened few"?

When we told pairs of adults either to win or to learn in arguments, we found that arguing to win mindsets resulted in their viewing one position or another as definitely true.[70] On the basis of a brief interaction, argue-to-win pairs viewed the truth as more black-and-white or objective. Those in the argue-to-learn group were more inclined to see some truth in the claims made by both sides; they realized that it is rare for one side to be perfectly right and another to be perfectly wrong. If a brief interaction could cause such an effect, imagine the stronger effect if personality, family, and culture guided children to take either a competitive or a collaborative approach toward argumentation.

Strict objectivist views of the truth are more likely to encourage unquestioning dogma, which is often associated with strong feelings of tribalism. Descriptions of some of the most dramatic advances of science often embrace the collaborative nature of discussions, and arguments, among members of a group. Nonetheless, sometimes, you should be competitive and argue to win. The legal profession is an obvious example. In most criminal trials, the prosecution and the defense have, and should have, the clear-cut goals of making a winning argument for their constituencies.

Does one form of argumentation dominate early on in development? Arguing to win might seem an early favorite as everyone has seen siblings heatedly insist that each is right.[71] Moreover, arguing to learn might seem to require more sophisticated views of the truth and of how knowledge in others is evaluated. However, young children seem quite capable of arguing to learn. Thus, the psychologist Michael Tomasello and colleagues showed that children as young as five could be influenced by "cooperative" or "competitive" contexts to argue in starkly different ways. Children were asked to reach agreement about where to place toy animals in a zoo. Each child owned half of the zoo. The cooperative context emphasized finding good homes for each animal and coming up with the best answer. The competitive context emphasized the goal of having more animals on one's own side and winning. Children in the cooperative context produced more reasons in their arguments as well as providing more reasons that considered both sides or both points of view. In addition, they preferred to engage with others who were willing to relinquish their views when given a good reason.[72] Researchers are just beginning to explore what sorts of situations prompt the two kinds of arguments,[73] but these recent studies suggest that before the start of schooling, children can be gently guided to argue in a manner much closer to what happens when a research group joyously makes a new discovery; indeed, all things equal, they may naturally prefer such forms of interaction. This may put them at an advantage over many adults in terms of the ability to learn from arguing.

Arguing to learn is a powerful early form of transactive wonder. When it is encouraged and supported, more elaborate cycles of conjecture and examination may flourish among groups and provide benefits going far beyond a simple dyadic argument. It offers a powerful antidote against competitive and destructive arguing-to-win mindsets that can arise from tribalism and zero-sum ways of thinking.[74] In many settings, arguing to learn can be an exciting and rewarding way to greatly expand the scope and depth of wondering.

■ ■ ■

By the time they enter their first classroom, most children have amassed a diverse and powerful set of techniques for learning from others. In almost all cases, these skills at tapping into other minds amplify the power and scope of their wondering. They engage in seemingly silly imitative behaviors that turn out to be effective ways to learn about complex devices. They grasp the implications of overt pedagogical demonstrations and use those implications for object exploration. They have surprisingly early abilities (in contrast to other facets of metacognition) to know when there are gaps in their own knowledge and to figure out who might be best to ask for help in filling those gaps. Social information-gathering contexts appear to promote metacognitive precocity. Preschoolers can judge what sorts of knowledge are plausibly learnable in different contexts ranging from social isolation to social immersion. Their portfolio of skills for evaluating the testimony of others includes assessing the intrinsic quality of statements (e.g., circularity and parsimony), linking the plausibility of statements to beliefs about plausible patterns in the world, and using consensus to guide evaluation. These abilities help them recognize the divisions of cognitive labor around them. They continue to refine their understandings of those divisions for many more years.

Given that many forms of information, both within and outside of the sciences, must necessarily be learned through social interactions, these abilities shown by young children should not be surprising. They are perhaps less obvious because we all underestimate how much we learn from others and overestimate our roles as lone individualistic learners. This lone-wolf myth is often mistakenly applied to many historical figures in science as well. A greater appreciation of these skills in young children should help us preserve and develop those skills and thereby supply children with vast new areas of wonder.

We also need to better understand children's goals in harvesting socially distributed knowledge. What is the target of their questioning, and what do they rely on most when they defer? They are especially attentive to causal information as indicative of expertise, but children sense something more than mere causal relations. They seem to be learning about how causal relations compose coherent mechanisms that can help delineate areas of expertise.

Children's knowledge-foraging abilities, however, are fragile at first and are vulnerable to decay with disuse. For example, if children have few sustained interactions with other people, they may lose the ability to evaluate the quality of informants. Similarly, if children experience nothing but argue-to-win interactions, they may be more inclined to argue in the same manner and thereby miss out on all the benefits of arguing to learn.

More ominously, the social harvesting of causal explanations from others nosedives as children enter school. Even preschoolers show radical shifts in question asking when they go from their homes to their preschools. A four-year-old may ask how and why questions at less than one tenth of the rate shown an hour earlier while at home. Teachers are somehow signaling that they are taking charge of question asking and that their students should work primarily on answering those questions. While teachers may occasionally invite children to ask questions, there are very few classroom cases of "passages of intellectual search" in which a child and a teacher have an extended back-and-forth conversation involving many questions on both sides that gradually build up an informative explanation. Yet, those passages occur all the time in most homes, at least until the start of elementary school. Teachers take over question asking for many reasons, ranging from the practical challenges of having extended bespoke conversations with each child, to the ways in which schools mandate curricula and assessments, to misguided views by most adults about what kinds of content children at various ages can learn.

We need to better understand how all these factors can converge and crush the vigorous wondering we see in young children. Only then can we consider how to promote exhilarating lifelong voyages of exploration and discovery for everyone. We are custom built for such voyages and prepare for them beautifully in the preschool years only to be steered off course into doldrums of aimless demoralizing attempts to master brute facts unmoored from the fascinating causal architectures that give rise to them.

We have repeatedly seen hints of the special role of understanding how parts of things work together in systematic ways to stably produce effects—their mechanisms. Mechanisms may play an outsized role in what we choose to learn from others and in how we explore meaning in the world around us. We now turn to what mechanisms are and why children appreciate their privileged status as they wonder about how and why.

4 The Mechanistic Mind

If you ask a professional magician to perform at your three-year-old's birthday party, she may either politely decline or substitute a child-oriented act that contains very little "magic." Adults' sleight-of-hand tricks and other misdirections often fail to impress young children who are not as susceptible to quirks of mature attentional systems. Indeed, many child-centered magic shows often succeed primarily because of goofy costumes and cute animals. My son Marty remembers his first exposure to magic as not really being very magical at all. It came in the form of packaged kits, special cards, and corny tricks, all of which were plainly not magic but instead were contrived forms of deception. The deceptions rarely worked on friends but did result in highly amusing performances.

But if you *do* want to perform real magic tricks, pretend to violate core physics by having a real ball seem to disappear or pass through a solid. Young children will indeed be surprised. What happens next, however, is less studied but more important: when the show is over and if the props are left on the table, many young children will immediately try to look inside or behind the magic box to figure out what was inside that made the magic work. They are, in short, searching for a mechanism to explain away the mystery.

When scientists gain insight, they are often described discovering the mechanism underlying a phenomenon. Despite such common descriptions, mechanism has only recently become a major topic for philosophers of science. While scientists often focus on mechanism, science viewed from a more formal philosophical stance was usually discussed in terms of such practices as engaging in deduction, induction, and providing evidence. Yet, mechanism is now considered central to understanding how science actually works, especially in biology. Mechanism is also central to understanding young children's fascination and exploration of all that is around them. It may be the most powerful way that children's understanding can grow through wonder.

Characterizing Mechanistic Explanations

Across all the sciences, mechanistic explanations are especially prominent in biology. In the physical sciences, discussions of mechanisms occupy less attention.[1] This difference occurs because the most cognitively compelling mechanistic explanations are often mixtures of function and mechanism. For example, in describing the mechanisms responsible for cellular mitosis, biologists break down the process of cell division into functionally distinct phases (condensation, spindle formation, alignment, etc.) with different mechanistic explanations unpacking how each phase works. In contrast, when describing the causal processes underlying a volcanic eruption, such functional language is not as prominent—things just happen sequentially and not as a nested hierarchy of functions each with its own mechanistic explanation. The US Geological Survey explains volcanic eruptions as occurring when magma pushes through vents and fissures to the earth's surface.[2] Magma rises up because it is lighter than the solid rock all around it. Nowhere is function explicitly described or implied.

In this sense, many biological systems are more similar to artifacts than the nonliving natural world. With artifacts of any complexity, mechanistic explanations can involve hierarchies that decompose into nested sets of parts and descriptions of the causal interactions of those parts. For example, an explanation of how a refrigerator works starts with its overall cooling function and then decomposes that function into the interactions of functionally described parts such as the compressor, the thermostat, the condenser, and evaporator coils. These components are then decomposed further into their own components to explain how they work. Explanations of evolved biological systems often have a similar structure.[3]

We will consider prototypical mechanistic explanations to be organized hierarchically as shown in figure 4.1.[4]

Returning to refrigerators, the hierarchy has an overarching process, usually described in functional terms. Refrigerators keep their contents cool. The next level down is a set of isolatable components forming a causal chain that unfolds over time. For a refrigerator, the chain might start with

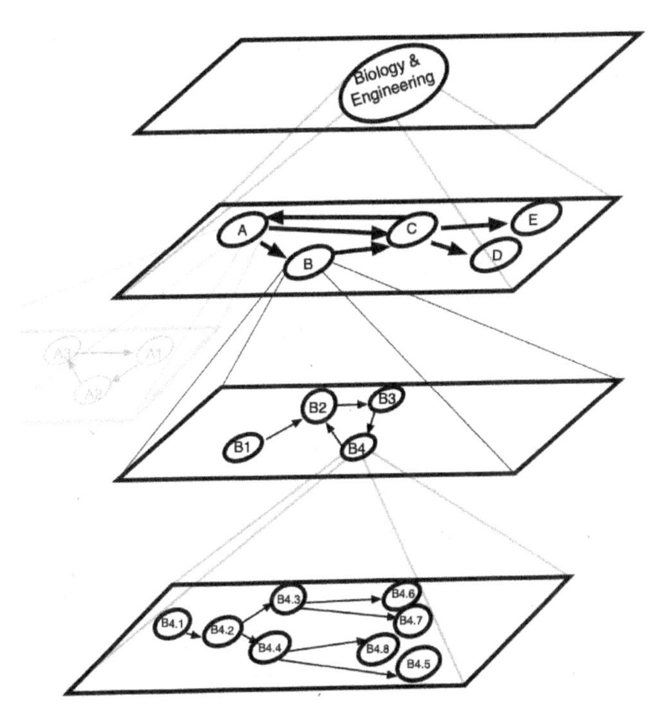

Figure 4.1

Typical mechanistic hierarchies in the biological sciences and engineering. Functions are explained by mechanisms and subfunctions that can, in principle, continue decomposing downwards for many levels. We see different kind of causal architectures, such as cycles and branching chains. Here, the unpacking of causal component B is shown. Comparable decompositions occur for A–E, with part of A's decomposition in light gray. In practice, practitioners in disciplines confine themselves to just a few layers, often just one. These hierarchies suggest clear starting and end points when, in fact, cycles are often closer to reality. A more accurate, but confusing illustration might use a three-dimensional cone. A strong grasp of mechanisms underlying an entity would involve a sense of this hierarchy and its major components and layers. (Inspired by C. Craver & J. Tabery, "Mechanisms in Science," in Edward N. Zalta, Ed., *Stanford Encyclopedia of Philosophy*, Summer 2019 ed., https://plato.stanford.edu/archives/sum2019/entries/science-mechanisms.)

the *compressor*, which functions to squeeze the refrigerant gas into *condenser coils*, which functions to turn the gas into a liquid and radiate heat through the coils, which then sends the liquid through an *expansion valve*, which functions to initiate changing the liquid back into gas form and to allow it to enter the *evaporator coils*, which absorb heat from inside the refrigerator. Further downward hierarchies decompose the compressor and other parts into components each with their own functions. The diagram suggests neat, discrete components organized into distinct levels with simple causal chains linking components at each level. The reality can be messier with complex interactions between levels and with causal cycles. In fact, refrigerator explanations also require a cycle since the evaporator coils feed into the compressor, but figure 4.1 is a more canonical illustration of how mechanisms are structured in biology and engineering.

For the nonbiological natural world, when scientists talk about mechanism, they refer to underlying causal patterns that afford insight into a phenomenon. Classical accounts of mechanism in physics often simply meant explanations purely in terms of causal interactions between physical material things without mention of function. In the realm of biology, however, analogies to constructed devices were offered to thwart the need for nonphysical notions such as vital spirits, goals, or free will. In 1651, the philosopher Thomas Hobbes proposed a mechanical view of the body in his introductory chapter of *Leviathan*:[5]

> For seeing life is but a motion of Limbs, the begining whereof is in some principall part within; why may we not say, that all Automata (Engines that move themselves by springs and wheeles as doth a watch) have an artificiall life? For what is the Heart, but a Spring; and the Nerves, but so many Strings; and the Joynts, but so many Wheeles, giving motion to the whole Body, such as was intended by the Artificer?

Similarly, Descartes discussed the clockwork mechanisms that could fully explain the material world (but not the world of the soul). These classical views of mechanism's role in the sciences were largely neglected in the philosophy of science until about two decades ago, when William Bechtel, Carl Craver, and other philosophers argued that mechanisms are central to understanding the success of science, especially in biology.[6] Since then, mechanistic explanations have become a topic of growing interest to psychologists. Early interest in mechanisms per se, not just in causal relations, suggests that they may have a privileged status in our searches for explanations. When children or an adults think "what if . . ." or "how could that possibly . . . ," they may have in mind a particular class of mechanisms. Mechanisms may naturally emerge in our thoughts as we try to clarify how and why questions.

Here, our focus is on mechanisms that include functional descriptions while also recognizing other function-free versions of mechanism. Even these accounts, however, go beyond a mere focus on physical causality. They typically describe structured causal relations of several elements, such as those organized in a hierarchy.[7] Mechanism is therefore much more than a mere list of causal relations. Critically, the new accounts of mechanism also suggest that mechanisms provide especially insightful ways of understanding the world. If that is the case, how does the appeal of mechanism arise in development? Does a focus on mechanism amplify the depth and reach of wonder by enabling more structured conjectures about what lies within and how such causal patterns might be broadly shared?

To address these questions, we examine four sets of studies: First, we ask if young children sense the special *epistemic*, or knowledge rich, value of mechanistic explanations. Then, we ask whether children grasp the specific ways that mechanistic knowledge provides unique knowledge benefits, focusing here on generalizability. Next, we explore children's interest in learning mechanistic details based on mere framings of the kind of information they could learn. Finally, we consider cases in which young children automatically learn abstract mechanistic patterns and show how these abilities suggested instructional interventions. Taken together, these studies illustrate how children's fascination with mechanism allows them to develop ways of thinking about causal patterns that enable wonder to consider whole new realms of possibilities.

Children Are Drawn to Mechanisms

Why might young children be able to appreciate the special merits of mechanism and seek it out in their inquiries about the world? After all, university professors constantly implore their students to specify the mechanisms underlying their vaguer claims. If asked to explain how classic mopeds restrict their speeds to the mandated thirty miles per hour, a student in an applied engineering course might reply, "By installing governors in them that made sure the moped couldn't go any faster than thirty miles per hour no matter how much you used the accelerator." Given such a response, a teacher would point out that the student merely restated the function and provided no new information on how governors actually worked. The teacher was hoping to hear how a centrifugal governor, when it rotated above a certain speed, disengaged the motor from the transmission. Hand-waving over mechanistic details or resorting to functional truisms are common ploys when you haven't learned the content. Despite these errors in college, young children not only seek out mechanism but they also grasp how functional and mechanistic relations are related to each other. They even sense how classes of mechanisms differ across broad domains such as animals, machines, and nonliving natural kinds. Well before they enter their first classroom, children appreciate the special merits of mechanistic insights and how they are organized in large-scale explanations. This early appreciation of mechanism is critical to understanding the rise of wonder early on, its subsequent decline, and its possible resurrection in later years.

Mechanistic explanations can be complicated and hard to remember in detail. Yet, very young children actively seek them out and appreciate how knowledge of mechanisms confers special benefits. Why are they so fascinated by mechanism when they, and most adults, retain almost no mechanistic details in either recall or recognition? The answer to this puzzle leads to a critical point: distilling mechanisms into cognitively useful abstractions is central to understanding scientific phenomena at any age and to leveraging knowledge in the minds of others.

Thinking and Learning about Mechanism

An early affinity toward mechanism provides a cognitive foothold for the growth and maintenance of wonder. Very young children love to investigate anomalies and reveal relentless drives to know what makes things happen. Almost fifty years ago, a laboratory study reproduced toddlers' behaviors at birthday parties with magicians.[8] Children watched an attractive toy being placed in a wooden box. The door to the box was closed briefly and then soon reopened. The old toy was gone and a new toy was in its place. Children as young as twelve months were described as puzzled and engaging in searches for causes of the disappearance of the original toy.

I vividly remember this study because it figured prominently in my first year in graduate school. If the children were puzzled by such disappearances, perhaps their degree of puzzlement could be used to measure similarities between toys at a conceptual level. It was an incredibly naive conjecture. I had been a biology major as an undergraduate, with little exposure to child development, and my ignorance showed. Measuring puzzlement turned out to be difficult. Because of that problem, in addition to many other flaws in the method, I abandoned the task after trying it out with a dozen children ranging from twelve to eighteen months.[9] However, in my preliminary tests, I did observe vigorous exploration and search behaviors for the cause of disappearance. I remember several children staring intently at the box and soon crawling around to the back side (from their point of view) to try to see if there was an explanation. They knocked on the box, lifted the door and peered inside, and pushed and pulled every part of the box they could. One child was so intent on exploration he held out his other arm with his palm outward indicating that I needed to back off and let him figure out what had caused the toy swap.

These largely preverbal children showed strong and persistent drives to figure out what unseen processes had caused the mystery event. Of course, because they couldn't speak, this interpretation was based only on inferences about their behavior. Just a few months later, children do start asking why and how questions, and those questions more directly reveal searches for mechanism. In one line of work, the psychologists Brandy Frazier, Susan Gelman, and Henry Wellman presented children at ages four and five with anomalous situations designed to trigger spontaneous questions.[10] For example, an actor might pour ketchup on a bowl of ice cream or a turtle might be shown lying comfortably in a bird's nest next to two baby birds. As soon as the children recognized the anomalies, they launched a surge of why and how questions, such as "Why is that turtle in the bird nest?" Despite occasional memorable exceptions, children do not normally ask why and how to annoy parents or to prolong social interactions; instead, these questions are triggered by noticing anomalies and trying to find a causal explanation for them. Their desire for a causal explanation was especially clear when the experimenters either answered the why and how questions with good causal reasons or answered with useless responses (such as "People don't usually pour ketchup on ice cream").

If the response did not supply a good causal reason, the children often reasked the original question, indicating that they hadn't received a good answer. They also were more likely to remember the causal responses than the useless ones.[11] Other studies on children's questions confirm the early presence of a special interest in learning more about the causal story behind observed phenomena and situations.[12] Studies of spontaneous speech in everyday conversations further show a steep rise in why and how questions shortly before most children go off to kindergarten.[13] It is as if children are excitedly honing their inquiry skills in anticipation of the mechanistic information they expect to be at the heart of their curriculum. Unfortunately, their school experiences don't usually turn out that way.

These studies on questions do not show a specific preference for mechanistic explanations as opposed to preferring causally grounded explanations in general. Some patterns of questioning were highly suggestive of a specific interest in mechanisms, but they were not conclusive evidence. Follow-up studies, including many from our lab, were needed to demonstrate a specific preference for mechanistic explanations.

Mechanism and the Quality of Knowledge

Sometimes a "mechanistic approach" means focusing on causation over simple correlations, contingencies, and other noncausal descriptions, but often more is meant. Great achievements in the sciences can reveal stable networks of causal relations. As research proceeds, that network may become ever more complicated; other times research moves toward an elegant idealized simplicity. Virtually every practicing scientist agrees that great explanations needn't specify *every* causal influence on the phenomena under study. In some cases, those influences, while present, just don't make interesting differences. Even though it theoretically exists, we rule out the possible influence of the moon's gravitational influence on eyeblink rate. In other cases, the influences might ultimately be massively important to the causal stability of a system, such as the quantum underpinnings of the Newtonian world of large physical objects, but going down that many levels of the mechanistic hierarchy to explain how oars propel a rowboat is going several bridges too far.

Assuming reasonable boundary guidelines on depth, are explanations focusing on mechanisms seen as indicating high-quality knowledge? In a series of studies, we focused on the perceived knowledge bearing value, or *epistemic status*, of mechanistic explanations.[14] Functional explanations, explanations based solely on the goals and functions of an entity or its parts, provided a salient alternative. While mechanistic explanations often include functional descriptions as guides to the layered components in a system, their focus is on the "clockwork" causal sequences linking together all the critical components. A perfect functional explanation of why a cow has four stomachs can simply state, "Because they help the cows eat grass." But, a decent mechanistic explanation must include the four stomachs and how their causal interactions unfold over time. (It is a quite bit odder than you might think.) We know from research by the psychologist Deborah Kelemen and others that children can find functional explanations especially appealing.[15] That appeal may lie more in the perceived value of functional information for various purposes as opposed to its real explanatory value,[16] but either way, it poses an alternative account of explanatory preferences. A different class of competing "explanations" might be ones that exclusively describe inside parts, but not their critical causal interactions, as in "A cow has four stomachs because it has four things inside its belly that are just like stomachs." A focus on insides has been linked to children having a bias for unseen inner essences. Or perhaps, attaching any causal descriptions to features will make an explanation appealing even if that list of descriptions lacks the coherence of mechanisms, as in "A cow has four stomachs because the stomachs cause burps, tummy aches, funny sounds, and hold food." Our studies teased these different interpretations apart.

We asked adults and children from ages five to eleven years to evaluate the knowledge of two people who provided different reasons for preferring one fictitious device or animal over another (fictitious cases ensured that familiarity with specific known examples wouldn't affect results). One set of reasons referred to mechanisms; the other represented one of the alternative forms listed earlier. The following example illustrates the kind of contrast employed.

Mechanistic reasons

Mr. Jones thinks the Seneca is the best electric car to buy because (1) The Seneca's strong engine lets you speed up quickly when driving; (2) The Seneca's motor runs very smoothly because it uses very slippery oil; (3) When you go on trips in the Seneca, the car's special springs keep it from bouncing too much on bumps; and (4) The Seneca's powerful lithium battery lasts for a long time.

Nonmechanistic reasons

Mr. Smith thinks the Seneca is the best electric car to buy because (1) The Seneca comes in so many great colors; (2) The outside of the Seneca is very shiny because of the new, special kind of metallic paint the builders use to finish it; (3) The seats inside the Seneca are made of soft Corinthian leather so they are extremely comfortable to sit in; and (4) The Seneca has lots of cup holders for your drinks.

After hearing the reasons, participants were asked to rate how much each character (e.g., Mr. Jones and Mr. Smith) really knew about the entity. Children of all ages saw the person providing mechanistic reasons as knowing more (Mr. Jones in this case). Even five-year-olds consistently saw the person providing mechanistic reasons as having deeper and more extensive knowledge than the other person.

Sensing Mechanism's Broader Benefits—The Case of Generalizability

Young children see those providing mechanistic reasons as having richer knowledge, but what makes that knowledge seem more extensive? Perhaps mechanisms are seen as widely shared among members of broad categories. We therefore asked adults and children from six to eleven to infer how some elements of knowledge implied others.[17] Suppose a person knows a great deal about how cars work. How much does that person know about trucks? Helicopters? Kangaroos? Suppose a person knows a great about how the parts of dogs work. How much does that person know about wolves? Cows? Grasshoppers? Lawn mowers?

We predicted that most adults would agree that the car expert probably knew a lot more than average about trucks, somewhat more about helicopters, and no more than anyone else about kangaroos. Similarly, the dog expert would be expected to know a lot about wolves, more than average about cows, somewhat more than average about grasshoppers and no more than others about lawn mowers. Now, contrast this first case with a person who knows many facts about cars or dogs, such as where the first car engines were built or how many different companies make tires for cars. For dogs, the fact expert might know how long it normally takes dogs to learn their names or the percentage of dogs that have floppy ears. We predicted less confidence and agreement about how much these people would know about other categories.

We told children that different pairs of identical twins each read a book with information about the same category of things. The categories were initially presented at the most commonly used level of categorization known as the *basic level* (e.g., clocks) and were illustrated with the same six images of instances, but the two books had very different kinds of information. One twin learned about how things in the category work, and the other twin learned facts about those same things. For example, we said that one twin learned "what makes the parts of the clock move . . . and how clocks can keep working for years without stopping," and that the other twin learned "where clocks were first invented . . . and how many clocks are made every year." In another case, we said one twin learned "how sparrows' stomachs break down hard seeds and insects . . . and how sparrows' voice boxes vibrate to make chirping sounds" and the other twin learned "how long it normally takes sparrows to search for food and at what time of day sparrows chirp the most." The stories described identical twins learning from books so as to control for any differences in background ability or knowledge. In addition to clocks and sparrows, we also tested cars, smartphones, sharks, and tigers.

After presenting what each member of the pair knew, we asked who knew more about various target categories at other levels of categorization, such as the superordinate level (e.g., machines), and an unrelated basic level category (e.g., tulips). For example, "Here are some machines (show six examples of machines). Who do you think knows more about machines? Adam who knows about how clocks work, Bob who knows facts about clocks, or do you think they know about the same amount?"

Children as young as six thought the mechanistic twin would know more about the superordinate category than the facts twin but would not know more about the unrelated category. Younger children showed this pattern for just the artifacts (e.g., clocks). Only older children and adults showed an equally strong preference for both artifacts and animals. This developmental difference may reflect the greater transparency of mechanisms for simple artifacts than for animals. Children are also more likely to observe others fixing artifacts as opposed to "fixing" animals. This contrast also maps roughly onto greater difficulties historically of understanding animals in purely mechanistic terms. Acquiring an elaborated understanding of animals and plants in mechanical terms may be one of the most dramatic potential effects of teaching science well during the early school years.

Mechanistic knowledge focusses on causal patterns central to understanding how members of a category work. Such patterns are rarely confined to tiny domains. This property about the world may drive intuitions about generalization. We may sense that stable patterns of this sort are hard to come by through either engineering or natural selection. Consequently, they will be widely shared with other categories in some larger superordinate kind. In contrast, mere facts don't have this feeling of needing to be broadly shared. Knowing the incidence of floppy eared dogs doesn't seem to imply much about what else is known about other animals. Mechanistic knowledge is not a simple list of statements. Mechanistic elements are highly interconnected through causal relations and form a stable structure that persists over time. Knowing one thing in this system well may imply at least some understanding of the system as a whole. We see glimmers of this pattern in our other studies and in related studies by other groups suggesting more powerful causal insights are associated with mechanistic knowledge.[18] When we wonder about possible mechanisms, we are also more likely to extend those possible models to much larger sets of things in the same broad category than if we merely wonder about a fact, such as color or size.

Anticipating the Promise of Mechanistic Framing

Why and how questions appear around age two and increase as much as tenfold by age five, and children strongly prefer mechanistic responses. Do those preferences persist even when children are simply asked about what they'd like to learn about but without details on the content? In a project with five- to ten-year-olds, we presented booklets with cover images of the same object but with different colored backgrounds indicating they were different books. We started to read from each book with a brief introductory sentence fragment that cued for different kinds of information while not providing any real content. The mechanistic booklet began with a phrase like "The way a Venus flytrap works is . . ." while the other booklet began with phrases that promised entertaining information, for example, began with a phrase like "Some surprising stories about Venus Flytrap are . . ." We then asked children which book they wanted to read next. Some children were told that their goal was to learn as much as they could and to pick the books that are the best for learning. The second group was told to pick the books that were the most fun. Finally, a third group was asked simply to choose the books they wanted to hear the most, a condition designed to elicit "default" preferences.

With a learning goal in mind, most children in all age groups preferred the mechanistic information. In the entertainment and the default conditions, children up to age eight generally showed roughly equal preferences for the two kinds of information. Somewhat older children seemed to prefer entertaining information, at first just in the fun condition and then, about a year or two later, also in the default condition. By age ten, children seemed shift *away* from seeking mechanistic information in the default condition. The spontaneous desire to learn more about the mechanisms underlying a diverse range of things (a flytrap plant, a submarine, a storm cloud, a firefly, or a helicopter) declined as they approached middle school, a trajectory many science education studies confirm.

Why does the intrinsic appeal of mechanistic information apparently decline in later elementary school? Perhaps older children think they know most everything due to illusions of knowing. Perhaps they construe "how it works" as containing more technical and boring information. Perhaps they simply have developed an aversion to mechanistic information of any kind. Yet, when asked whether they want additional information given intriguing statements about animals, children as old as ten are unsatisfied with circular explanations and ask for more information. In contrast, they accept mechanistic responses.[19] Thus, while spontaneously seeking out fewer mechanistic explanations, older children still clearly prefer them when pitted against circular ones.

Sensing the Fit between Functional and Mechanistic Explanations

Young children believe that someone who knows an object's mechanism generally knows more than one who knows its function. But do they also understand how functional and mechanistic information might best be combined in instruction? Adults favor explanations that start with a brief functional framing followed by a mechanistic elaboration. An explanation of how a GPS works should start with a comment on its general function. This helps orient the learner to the mechanistic explanations that follow. Functional information provides guidance about what mechanistic details will be most relevant. If an explanation starts with mechanistic information, it may be difficult to know how to prioritize additional information as it flows in.

Mechanistic and functional explanations work together in ways that we are just beginning to understand. The most prolific researcher on the topic, the psychologist Tania Lombrozo, has shown that adults generalize properties differently depending on whether they are explained in functional or mechanistic terms. When a property, such as skin color, is explained mechanistically in terms of causal descriptions of its component parts and processes, people tend to generalize other properties of that animal to new animals that share the same parts and processes. In contrast, when skin color is explained functionally (e.g., it attracts insects so as to help pollination), people tend to generalize other properties of that animal to new animals that share the same function.[20] Functional explanations also pose lower cognitive loads and may come to mind more easily when people are distracted or otherwise cognitively challenged.[21] Because functional explanations can be compatible with many different possible mechanistic systems (think of all the mechanistic ways the function of carrying an object by air travel can be implemented), they tend to require far fewer details and can avoid describing the cause-effect chains that give rise to the properties and behaviors of a thing.[22]

Sophisticated mechanistic mindsets, especially for biology and technology, will always involve an interweaving of function and mechanism at several levels. Young children, while showing some sensitivity to the relations between these two modes of explanation, have the potential to greatly elaborate on such relations during the school years. But this will only happen if they receive many detailed examples of both modes working together. As illustrated in figure 4.1, function and mechanism interact in highly specific and mutually informative ways. When children begin to see the power of such structured relations in the early school years, their abilities to learn and create their own explanations and conjectures will gain a huge boost, but this surge may only occur when these structures are embraced and employed by those around them.

Abstraction from Detailed Mechanisms

Why should children be so interested in mechanistic information early on? If its value was in accurately allowing them to remember all the details, their interest should be short-lived. Explain to children how virtually anything works in detail, and they will later recall little of what they heard. Why don't they soon learn not to bother paying attention to mechanistic information given how unlikely they are to recall it?

Complexity as Mechanism Metadata

Children's interest in mechanism is justified not because of detailed recall but because they can infer from intricate mechanistic details more abstract generalizations of great value. For example, they can acquire an enduring sense of the causal complexity of a system even as the details quickly evaporate. Imagine hearing explanations of how a mousetrap and printing press work. The mousetrap has only a few parts, and you envision a simple system that would be easy to figure out. In contrast, a rapid one-time detailed explanation of how a printing machine works quickly overwhelms the listener. Causal complexity is different from visual complexity or mere part counts. Surface decorations can be visually complex but causally irrelevant. Some devices with tens of thousands of parts, even moving parts, are much simpler than other devices doing the same thing with fewer parts; consider an hourglass compared with a simple mechanical clock. Causal complexity requires an appreciation of more subtle causal relational patterns such as the presence of structured causal chains and hierarchies. It also includes a sense of the heterogeneity of parts and processes. We described such information about causal complexity as mechanism *metadata*. Metadata traditionally refers to information in digital image files that summarizes image properties, such as file size in memory. Similarly, judgments of complexity summarize information abstracted from the details of a mechanistic explanation.

Do young children have stable senses of causal complexity, and what information do they use to infer complexity? We asked children and adults the causal complexity of devices and biological components.[23] These devices ranged from flashlights to submarines and from hair to eyes. Starting at around age seven, participants largely agreed on the complexity of both devices and biological components. Even at five years of age, children's complexity judgments roughly aligned with that of older children and adults, although they exhibited more variation. The greater variability may reflect larger differences in personal experiences with the objects. Those experiences become more similar as children accumulate shared experiences in elementary schools. Critically, all ages used their complexity intuitions to guide judgments of when to seek help in learning about something or in fixing it—more complexity meant greater help-seeking. Nonetheless, all ages made these judgments while knowing hardly any details about the actual mechanisms.

Complexity intuitions do change over the school years. In another project, we found children becoming more adept at using different kinds of cues to infer the mechanistic complexity inherent in devices. First, we asked if behavioral diversity implied more internal complexity.[24] If one machine picks daffodils and daisies while another picks daffodils and strawberries, which has more complicated insides? Adults prefer the daffodil/strawberry picker and often explain how the actions for picking those two things are more different from each other and would require more complicated configurations of internal parts to make both possible. Adults did not know how those parts differed in detail; they just immediately sensed that one was more mechanistically complicated.

In stark contrast, children at ages four and five, while knowing clearly that the actions of the strawberry/daffodil picker were more different from each other than the strawberry/daisy picker, could not make the next inference about internal causal complexity. They did infer that a machine that performed two functions (e.g., picked strawberries and daffodils) was more complex than a machine that picked just one (picked just strawberries), but not until age six did they begin to grasp how complexity was related to functional diversity. While this ability to use diversity as a cue first appears around the first grade, it is never explicitly taught. It emerges automatically without any guidance, presumably from encountering a sufficient number of real-world cases that illustrate the pattern.

In other studies, we explored how children link interior configurations to object complexity.[25] By age five children start to match complex insides to artifacts they considered complex. These results fit with studies where preschoolers assumed that novel devices with more variable behaviors were linked to interiors that were globally more complex.[26] At first, children may rely mostly on how many distinct parts are present. Later, they consider part heterogeneity and part interconnectivity structure as cues to complexity. These three features are only a subset of all the interior patterns that we ultimately link to mechanistic complexity. Again, these relations are never explicitly taught to young children. They may be inferred from directly viewing the insides of various machines and linking them to how the machine is used. Learning from direct experience would also surely involve generalizations to other presumably related categories (e.g., insides of car engines to insides of truck engines). Learning indirectly may occur by hearing others talk about the insides of various things and how difficult they are to fix or put together.

When young children encounter a device of great complexity, they soon lose track of many details but at the same time are forming enduring impressions of overall complexity of the device. They also extend these complexity inferences to other devices assumed to be in conceptually similar categories. This complexity knowledge then helps a child know when to seek expert help and how much expertise is needed. Complexity judgments are just one kind of metadata arising from exposure to mechanisms. Other useful abstractions could include whether the insides all work at the same speed; whether the parts move linearly or rotationally in the same direction and at the same time; whether the insides undergo irreversible changes in performing the function; whether the device amplifies or reduces such features such speed, force, energy, or heat; whether the device is fully mechanical or also electrical and chemical; how often the system fails; the precision with which the device works; and whether it self-adjusts to environmental changes. In addition, multiple interactions of two or more features might be critical, such as that precision improves or declines with increased use.

These examples illustrate the many useful abstractions we could learn over a lifetime. Related questions arise concerning how much of this can happen outside our awareness and simply through exposure. Are some forms of abstraction especially difficult to explicitly teach? For example, one can form an idealization that looks nothing like the instances encountered (e.g., a flowchart of subcomponent sequences) or a gross simplification that preserves overall visual similarity (e.g., a highly simplified illustration of a jet engine). Ultimately, we would like to know how and why children and adults exclude some types of information in the original and keep others in forming abstractions. We would also like to know how much detail is helpful in building up these abstractions even though it may soon be forgotten. Insights here are critical to understanding how we learn to read maps, use the desktop metaphor on computers, and draw informative diagrams. To improve cognitive efficiency, we must constantly abstract away from details.

Learning New Abstractions by Teaching Detailed Mechanisms

Mechanistic details provide conduits to more abstract representations, even as those details are largely forgotten. This process normally happens over weeks, months, and years of cumulative everyday experiences. Could it also occur during one brief lesson? We first asked this question in an elementary school with test scores placing its students far behind peers in science achievement.[27] For about fifteen minutes, using detailed diagrams, we explained in detail to children in grades 1–4 how a cylinder lock and a human heart worked. Some students heard a detailed causal mechanistic explanation of how each worked. Other students heard an equally detailed nonmechanistic description of the same parts, such as their typical color and location in the larger context (e.g., in the body or the doorframe). Prior to any descriptions, we asked all children to rate the mechanistic complexity of the heart and the lock. Then, right after the description, we reminded them of their initial complexity judgment and asked if they wanted to keep their answer or if the heart/lock was more complicated or simpler than previously thought. Finally, three weeks later, we asked them to recall mechanistic details and to provide one more complexity rating. We also asked them to rate a novel object that shared a larger category of body organs with the heart, a kidney.

Children who received mechanistic explanations and who initially judged the heart or lock as simple gave higher complexity ratings compared with children who received nonmechanistic information. The effect also was found again after three weeks. Mechanism children also rated the kidney as more complex than nonmechanism children. Revised complexity intuitions therefore generalized to a larger domain (body organs). After three weeks, the new complexity ratings endured while recall of mechanistic details dropped. Virtually all children were fascinated by the explanations even though the descriptions were far more complicated in terms of causal details than any science materials they had encountered in their school. They changed their mechanistic complexity ratings even though we never mentioned complexity or related concepts.

The study with hearts and locks raised the question of whether even more complex mechanistic information could instill abstractions. In one study, we used an animation of how a four-cylinder internal combustion engine works. This seven-minute technical video was designed for training adult engine technicians. It showed thousands of distinct causal events. The only major change from the adult version was to simplify the voice-over vocabulary and the grammar. When we asked lay adults, including many parents, whether the video would be instructionally useful for schoolchildren, they overwhelming said it would go completely over their heads and that elementary schoolers would quickly tune out.[28]

To explore abstractions other than complexity, we focused on three themes that are unfamiliar even to most adults: synchronicity, decentralized control, and containment. None of these ideas were explicitly mentioned in the video, but over time, it implicitly reveals that parts must work in synchrony, that the sequence of events is not choreographed centrally, and that a tight containment of gas must occur for a brief moment. We showed the video individually to children between five and eleven years of age and tested their ability to judge which of two alleged experts was the real car engine expert. The real expert stressed the importance of one of three themes, while the other expert stressed the opposite theme (that some parts went slower when others went faster, that there was a central controller, and that the gas must never be completely enclosed).

Even the youngest children exposed to mechanistic information chose the real experts more often than those who did not receive mechanistic information. The children also greatly enjoyed watching the videos. Only one child looked away in boredom and indicated a lack of interest. Many caregivers, who were present, were astonished at both their children's sustained interest and at their successful learning. More generally, we have in found in other studies that adults, including veteran teachers, greatly underestimate young children's ability to learn about mechanisms. Adults also greatly underestimate children's intrinsic interest in mechanisms.[29]

Children's natural ability to develop abstract understandings through exposure to mechanistic details can be either reinforced or discouraged by describing specific learning goals. In a study on the effects of describing learning goals, we edited an adult instructional video about the human heart down to four minutes and added a child-oriented narration. We contrasted learning when a child was given either the goal of learning how a heart works (mechanism goal) or the goal of learning the names of all the parts of the heart (labels goal). Children were then assessed on their ability to detect incorrect diagrams of blood flow and to decide what part of the heart was broken given certain problems with blood flow. Children as young as six performed much better with the mechanism goal. They not only were better at learning subtle principles concerning the workings of the heart but also at learning the names of heart parts. A better grasp of mechanism seemed to support name learning by making those names more meaningful.[30] A mechanism mindset helps learning of additional non-mechanistic content, while a label learning mindset results in relatively worse learning of the explicit target content. Despite this, elementary school instruction and tests often focus heavily on learning what things are called and not on how they work.

In retrospect, successful learning from these complex displays should not be surprising. Before the advent of child-oriented instructional materials, the only "curriculum" consisted of real-world experiences supplemented by scaffolding and attentional guidance from others. Simply learning how to get home from a neighbor's house requires abstraction from millions of details in successive scenes to an economical mental map. Similarly, children seem to be inferring causal "maps" that contain information about objects in terms of their complexity and associated causally central patterns. We needn't worry that they will be overwhelmed by the details of mechanism when the cover is removed on an operating vending machine or an escalator. They have surely evolved ways of recovering useful information from events whose complete contents cannot be tracked by any one person. This is one of the central themes of research in adult perception and cognition. We all must find ways of simplifying the world while still recovering useful generalizations. As this skill is rarely explicitly taught, it makes sense that it emerges early in rudimentary forms.

The Central Importance of Mechanistic Abstraction

A mechanistic mindset includes a precocious and spontaneous tendency to infer abstract causal patterns from concrete details. Thus, exposure to mechanism often has an ironic result: details are forgotten for the sake of creating abstractions. Details may be critical to generating abstractions in the absence of explicit instruction, but they are often discarded to yield useful abstract causal schema. Exposure to details may enable learners to detect patterns that are difficult to verbalize. This process may help them construct interpretations out of the details rather than simply memorizing a taught rule or principle.[31] The abstractions can then be deployed to make sense of a wide range of new situations. They also facilitate relearning of old material by helping relearners zero in on the most relevant parts of a lecture or reading.[32]

Many questions remain concerning how abstraction works, how causal patterns are ultimately represented in the mind, and how they are deployed. They can be construed as parts of a conceptual tool kit consisting of different abstract causal schemas. Those schemas provide insights into broad domains. The education scholar David Perkins invoked a similar notion called "mindware," described as schematic abstractions serving as useful knowledge tools.[33] For example, a child might notice from glimpses of the inner workings of a piano that discrete events occur inside corresponding to each finger press on a key. From these observations, she might form an impression of how musical instruments make rapid onset sounds. Brief physical contact must happen between two things, such as a hammer striking a string, a finger plucking a string or a stick hitting a drum. Perhaps this idea is generalized further to a class of signaling systems such as fire alarms and doorbells.

Abstraction enables individuals to assess the plausibility of attempting certain tasks. If abstractions reveal high complexity, long time spans, or action at great distances, they may lead people to quickly conclude that a project can't be done on one's own. Abstractions allow us to more easily generalize to related artifacts and biological systems and not be bound to concrete particulars. One might learn that cycles are central to many biological systems and anticipate them when exploring new cases. Abstractions allow us to guess who are likely experts. If someone has mastery of a phenomenon that we know is strongly linked to an abstract causal pattern, we suspect that person is likely an expert on other topics involving the same patterns, and indeed, children make such judgments all the time.[34] Abstractions allow us to combat dogma and misinformation by looking beyond superficial relations and similarities to see if anything of substance lies behind an invidious comparison. Finally, they give wings to wonder, allowing us to soar over a far larger range of possible worlds.

Fixing, Adapting, Improving, and Healing—Intervening on Mechanisms

Young children are fascinated by mechanism and go to great lengths to peer inside things and take them apart just to learn how they work. While these activities are driven by a desire to simply know and understand, children may also want to fix, adapt, improve or heal a device. When we intervene on mechanisms so as to change them, we often uncover otherwise hidden causal patterns.[35]

The first interventions on mechanisms by humans may have occurred with the growth of tool use. The first tools were things like chipped rocks and sharpened sticks with no decomposable parts. More complex devices made out of different material components, such as the bow and arrow, first appeared around 75,000 years ago.[36] Early tools may have put a selective pressure on brain development that enabled the creation of more complex multicomponent devices with discrete subfunctions.[37] Emerging cultures that supported the social transmission of knowledge may have also greatly accelerated the growth of technology.[38] As a result, the world of 35,000 years ago contained many multipart devices with identifiable subfunctions such as notches for holding bow strings or feathers for stabilizing arrows. By 10,000 years ago, relatively complex devices were present in all areas of life such as shelters and clothing, agriculture, and hunting.

These more active object-directed interventions have long been commonplace in young children's lives. In traditional cultures, children are deeply embedded in such actions when they help caregivers do chores and when they serve as apprentices. Many children in my generation were frequently asked to help fix things such as toasters, bicycles, and simple lawn mowers. These activities offered unique insights into how devices worked. In contrast, attempts to heal sick animals and plants didn't usually yield equally useful insights into the mechanisms supporting the life of living kinds.

Could earlier developing understandings of tools support later developing mechanistic understandings in biology? In our lab, we have seen hints of mechanism being grasped earlier for devices than for living things. For example, children seem to appreciate how function and mechanism work together in explanations of devices before they have similar realizations for living things. But those were preliminary incidental findings. Our studies on the division of cognitive labor also showed that young children's disciplines of expertise shift in ways that fit with an early emphasis on fixing, healing, adapting, and improving activities. As children come to understand the division of cognitive labor, they shift from seeing knowledge as clustered by occupations (healing, farming, fishing, carpentry, teaching) to knowledge clustered by stable causal patterns in the world (biology, physical mechanics, psychology, etc.).[39] But we still need to conduct studies on how actions such as fixing might enhance mechanistic understanding and wonder.

What implications should we draw from the study of ancient humans? Did children more than 100,000 years ago have no drive to spontaneously wonder about the world because whatever launched the explosion of tool design was not yet present? Or were they actually able to discern mechanism in nature long before the emergence of human-created mechanisms? Perhaps an early intrinsic fascination with mechanisms in nature was present without the distractions of machines. While some mechanisms in biology are invisible to the naked eye, others are easily observed. A child might infer the functions and mechanisms of certain parts of the eyes, such as eyelids, lashes, tears, pupils, and even blood vessels. The differences between eyes in predators and prey can lead to other mechanistic inferences and wondering. We may never know from physical traces how mechanistic understanding first emerged in ancient humans, but future studies of children's earliest appreciations of mechanisms in the natural and artificial worlds may shed light on our remote ancestors as well. Because contemporary cultures can vary greatly in what causal patterns they notice in complex biological systems, such differences may provide revealing developmental comparisons. For example, conceptual contrasts often arise from seeing ourselves as part of nature or as mastering and managing it. Many Western educated people tend toward mastery/management perspectives and may therefore be less sensitive to the first ways that early humans might have appreciated mechanisms.[40]

These considerations of the historical and developmental origins of mechanistic thought raise another set of questions about whether the world of today is changing rapidly in terms of how mechanisms are encountered by children and adults. We now consider this topic from a broader perspective of what is changing and what it might mean.

An Expanding Mechanism Desert?

This chapter, and the two before it, describe how the early years equip children with a rapidly growing ability to engage in wonder. They detect causal patterns of startling complexity and abstraction, they deftly leverage the contents of other minds, and they appreciate the power of mechanistic explanations. Their cognitive futures seem bright when these abilities support their intrinsic fascination with how things work. In the chapters that follow, we will rapidly descend from this optimistic perch as we see all the ways things can go wrong. To brace us for what is coming, we briefly consider here one way the gifts of mechanistic insight may be increasingly thwarted because of changes in the nature of technology.

Over the past hundred years, the Sahara Desert has increased dramatically in size, expanding by more than ten percent in area.[41] Formerly green nondesert regions have become expanses of arid sand, presumably yet another consequence of global warming. Has a comparable disappearance of rich resources occurred in our everyday experiences of mechanisms?

My first car was a used 1963 Triumph Spitfire. Because my car had no radio, it didn't have a single transistor in the entire car (see figure 4.2). It had a simple wiring harness and moving metal parts. The complete repair manual was a thin, heavily illustrated handbook that could be easily read in a long afternoon sitting next to a broken car. Because I had little money and an unreliable car, I spent many hours locating parts in junkyards and working on the car. Like many others in my generation, everyone assumed

Figure 4.2
Increasing mechanistic opacity. A 1963 Triumph Spitfire (top) had no transistors at all if the radio was omitted. In contrast, comparable 2021 cars, such as a Mazda Miata, have several billion transistors. While many mechanical operations were easily seen under the hood of the Spitfire, the Miata's workings are mostly obscured.

that you could fix your own car if you needed to. For many of us, working on cars was one of our most common activities after sports and school.

In 2021, most new cars have over ten billion transistors. The repair and service manuals cover how to use code displays that show what is broken. The manuals then tell the mechanic how to remove the broken part and replace it. Each removable part may have many parts of its own that are embedded in impossible-to-service enclosures. There is no book with an exhaustive description of all the car's parts along with diagrams of how they fit together. The average teenager today might know how to adjust the car seat or possibly how to replace a taillight. I recently asked some undergraduates what they could fix or service on cars. A surprising number said they couldn't do anything other than drive them and learn how to operate some of the commands on the touch screen. A small minority thought they could change a tire safely or use jumper cables correctly, and just one said they'd feel comfortable changing the oil. In contrast, in 1975, when our Chevy Vega engine died, my wife's brother pulled the dead engine using a rope and pulleys we rigged under a deck and then popped in a new engine. He did it all without a manual. He simply looked at the attachment points and figured out what needed to be done. He also was not an auto mechanic (he was a computer software specialist) but had "messed round" enough with cars to be able to understand and take apart any car he encountered.

The change over the past fifty years is enormous. Throughout my childhood and adolescence, I was often asked to fix toasters, floor lamps, simple clocks, record turntables, lawn mowers, and many other common household items. The default assumption was that you could take the cover off almost anything, peer inside, and figure out was wrong. Sometimes, the sheer number of parts was daunting or they were too small and/or fragile to work on as a novice, but even in those cases it was usually clear what the experts were doing. Today, very few things easily reveal how they work. When I recently tried to repair a budget toaster, I was surprised to find a circuit board on the bottom. The toasters of my youth had simple wiring and a bimetallic strip that bent at different temperatures so as to close or open a circuit. A fully transparent set of causal steps in older toasters had morphed into an undecipherable and unserviceable circuit board.

Repair shops for various appliances have evaporated. In the past, even small towns might have multiple television repair shops. Today, almost no such shops exist, and when they do, they usually mean they can replace modules on very expensive high-end TV systems. No one seeks local repair of the 32-inch TV that they just bought at a big box store for $85. They either get a replacement under warranty or buy a new one. This new reality has led to the "right to repair" movement, which argues that the inability to repair home appliances is a violation of consumers' basic rights.[42] Legislatures around the world pass bills asking for all or some of the following to be declared as rights: replacement parts, tools needed for repair, diagnostic software, and manuals or other guides. All of these must be available at reasonable prices. Notably, the movement has given up on the right to understand how the things work; if the manufacturer provides a way to follow a menu and repair the device inexpensively, the rights have been honored. In the end, this may not be enough. People also have a right to know how the things they own work—perhaps not in complete detail, but well enough to understand the critical components, how they interact, and why they are there.

My first few cameras were decipherable physical mechanical devices. It was easy to see all the physical parts and the clever ways the film was advanced. The chemistry was also straightforward, and many of us easily learned how to make prints in a darkroom. Todays' digital cameras are vastly easier to use and are much more versatile, but their workings are a mystery. Even worse, cameras as such have largely disappeared as almost all photos are taken with cell phones. Some of the image processing is now centrally shared with other phone functions, making it even harder to figure them out.

Few devices have escaped the trend toward increasing opacity of mechanism. The trend is compounded by additions of gratuitous electronics to things that worked fine without batteries and wires. Most greeting cards at our nearby drugstore now have electric components that play songs, speak greetings, make odd sounds, or flash lights. Most of the toys in Walmart need batteries. Even some toy hammers need batteries. Hammers would seem to be one thing that couldn't possibly have an electric component, but toy makers found a way. The electronics supposedly helps babies learn about hammers.

When electrical circuits were around us years ago, they were also more "mechanistically" transparent. In my elementary school, everyone was given a box full of wires and a handful of metal pieces and told to follow the instructions to build a working electric motor. We quickly learned what each part did and how it fit with others. Similarly, we all built crystal radios. To this day, I marvel that you can build a tunable radio from "raw" materials that works without a battery. The closeness to raw materials was important. With just copper wire and the "crystal" and a simple pair of headphones, you could see how radio waves were transformed into vibrations on a membrane that made sound waves. The idea that the radio set was powered by the inherent energy in the radio waves all around us still amazes me.

Shop classes, where one might learn how to build and repair things, have largely disappeared from the public schools.[43] When there are movements to restore them, the justifications voice concerns about educating those of our students who should go on in vocational work and the "trades." But that wasn't how those classes were perceived years ago. They were often considered a basic knowledge vocabulary and conceptual tool kit that everyone should share. Actually, it was usually every boy. The girls were sent to home economics classes instead. This inequitable division along gender lines was "solved" by canceling those classes rather than opening them up to everyone.[44]

One response to this concern is to praise today's youth as geniuses at computers and coding. These "digital natives" have knowledge that most of us adults can barely grasp. But this new knowledge seems unlike any version of mechanistic understanding. To be sure, our three sons, and much younger children, are wizards with computers in ways that I will never know. In a few minutes they are immersed in novel video games while I am still trying to move my agent. They figure out how to use new apps or software without reading any instructions and can seamlessly blend many software programs together to accomplish a task. I surrender completely to their greater abilities in this area. But what have they mastered? If today's computers were cars, have they learned how they work or have they merely learned how to drive them? They seem to have become experts at learning user interfaces, not at how the underlying code or physical components work.

Perhaps twenty-first-century children are learning a higher level of "mechanism" covering how programs and apps causally interact, but that isn't how they usually talk about their skills. Their knowledge resembles procedural routines more than programming skills. If you ask a teenager to give examples of how loops, queues, stacks, and arrays interact in simple programs, you may be met with a blank and sometimes hostile stare. Most teenagers' coding skills don't seem to be symbolic equivalents of advanced mechanical understanding. If they were, companies would not be madly scrambling to find even marginally competent programmers. Today's youth, the so-called digital natives, often seem to be gliding along on the surface of a vast system that they do not understand at all. They may have remarkable "native" skills at using and learning to use software-intensive interfaces, but if the real programmers have done their jobs well, the actual workings of those devices may be not at all transparent to those mastering the user interface. Software-shuffling skills at the user level may be completely different from knowing how and why. Teens may know that graphics processing unit chips are better for machine learning but not how they speed up processing. These gaps in understanding will only get worse as deep learning systems solve more and more problems for us. If a company comes up with an improved facial recognition system, people are unlikely to ask how its mechanism has changed. Even experts in the area just assume it arises from a more powerful deep learning system, which is a black box resistant to any useful interpretable explanation of how it works.[45]

In short, the world of artifacts has been radically transformed in the past fifty years. Readily apparent underlying mechanisms have vanished into mazes of circuits and blocks of silicon with millions, and often billions, of transistors. I remember when my father brought home a ten-transistor portable radio, describing it as a huge technology leap over our prior six-transistor one. It was even possible to know what each transistor did and how the four new ones improved performance. No one attempts such mechanistic accounts with the billions of transistors in current cell phones. Most young people have had far less daily exposure to how the parts of things causally interact to produce effects. Perhaps today's youth really are cognitively compensating with some version of explanatory knowledge at an abstract level of software functioning invisible to me. I would be overjoyed to see new waves of research documenting such knowledge. But even if it does exist, it doesn't justify neglecting the foundational processes that make those high-level software systems possible.

But perhaps my concerns are yet another example of an old codger moaning about intellectually lazy youth. Prior claims that radios, televisions, calculators, and the internet would make our children stupid all seem ridiculous in hindsight. I acknowledge that I may be just as mistaken. Perhaps this new generation has wisely outsourced their knowledge and understanding to the internet and the power of search engines. Why clog your head up with knowledge of how things work when you can simply Google the problem and watch a YouTube video to fix it? This argument, however, assumes that nothing is lost by outsourcing.

Relegating knowledge to search engines and apps differs greatly from actually knowing it. Outsourced knowledge can become unavailable when a device crashes or the internet fails. But even when all external sources are working fine, you think differently. If you cognitively know in your own head a rich array of causal patterns, you can problem solve and create new ideas more rapidly and fluidly. Our weekly lab meetings include many back-and-forth exchanges between about a dozen people. These exchanges require fast recall of models, patterns, and paradigms to recognize flaws or generate new ideas and solutions. If everyone had to pause and Google information, the process would take much longer or would not work at all. Moreover, without enough internal knowledge, you wouldn't know how to search effectively. People grossly underestimate how much prior understanding is needed to perform efficient searches for information.

If these worries about changes in technology have any validity, it will be ever more important to maintain an awareness of the encroaching mechanistic desert. Through such an awareness, we may find ways to preserve the benefits of a mechanistic mindset and how it empowers wonder. Indeed, some cognitive reforestation of technology may be essential to any practical steps we might take to nourish our wondering ways.

■ ■ ■

From an early age, children sense the special importance of mechanisms. They see people with mechanistic information as having deeper, richer, and more generalizable knowledge. They attend to detailed mechanisms and use that information to construct abstract interpretative causal schema. They then deploy the abstractions to interpret, understand, and explain novel phenomena. Far more than just noticing frequencies, correlations, and isolated causal relations, mechanistically minded agents build structured causal interpretations. These systems connect mechanism and function together in layered complexes. Those complexes are often at the heart of major intellectual achievements in the sciences, engineering, and other areas.

Mechanisms provide pathways for enduring higher order representations of how things work even as details rapidly decay. Young children, as well as adults, cannot retain mechanistic details of natural phenomena and devices. Yet, if teachers have misconstrued the nature of children's minds and concepts, retention of mechanistic details may be envisioned as the ultimate goal of learning. Young children may have a strong interest in mechanisms, but that interest serves as a pathway to learning more useful and robust forms of knowledge. Those forms include "mechanism tool kits" that provide insights into the kinds of mechanisms at work in a system without the user becoming overwhelmed. A child might learn that inanimate objects have nearly instantaneous effects on other inanimate objects (billiard-ball collisions) in contrast to events involving animate entities. A child might expect ropes and strings to pull or hold things together, or that containers with small holes or grids (e.g., colanders or nets) usually operate on fluids or fine aggregates (e.g., water or sand), or that springs can "store" a pushing force. These tool kits are often deployed "on the fly" and are used to construct tailored interpretations when there is additional support provided by the actual objects at hand as well as by other people. The right kinds of exposure to causal mechanistic information is essential to early science education. Rich mechanistic content cannot be omitted even though its details are forgotten. When kindergartners view surprisingly complex videos of the working insides of internal combustion engines, they acquire powerful abstract generalizations.

When we lose track of the pleasures of learning mechanism, we are cognitively deprived. A lifelong mechanistic mindset allows us to understand and engage with many communities of experts while arming us with the means for not being misled by others. In addition to simply using mechanisms for the pleasure of expanding understanding, we can use mechanistic patterns of inquiry to debunk fake experts, even as we know little ourselves. Frauds find it easier to make up fake facts and cite sources than to coherently provide more mechanistic details to a novice questioner with a mechanistic mindset.

Twenty-five years ago, the columnist William Safire nicely captured the joy of knowing mechanism as follows: "I think we have a need to know what we do not need to know. What we don't need to know for achievement, we need to know for our pleasure. Knowing how things work is the basis for appreciation, and is thus a source of civilized delight."[46] We all have this interrogative ability within us, but if we fail to maintain it, we can become victims of disinformation. When formal education, informal venues, and the culture at large clashes with our early emerging natural ways of embracing mechanism, we cripple wonder as well. The extraordinary rate at which preschool children develop their understandings of the world through wonder should not plateau and then dive when they enter school, yet, for many, that fate awaits.

II The Big Sleep: Weakening Wonder

5 Developmental Disconnects

In 1974, as a graduate student, I conducted a study in a small school in the foothills of the California coastal mountains. The study explored how schoolchildren perceived different types of ambiguities in language and vision. For example, it asked whether children's abilities to detect structural ambiguities in language (e.g., the two meanings of "the happy captain's wife") were more related to abilities to detect structural ambiguities in images (e.g., seeing either two faces or a single vase in an image) or to abilities to detect symbolic ambiguities in language (e.g., two meanings of the word *bank*).[1] One of the kindergarten teachers asked about the study, and, after listening carefully and looking at the materials, she confidently predicted how it would turn out. She declared that the outcome was obvious because the kindergartners were still in the concrete stage of development and would behave very differently from fourth graders who had transitioned to abstract thought. Forty-four years later, I was visiting a science museum where a staff member explained how a new exhibit had some features designed to appeal to younger children's concrete ways of thinking and other features designed for older children's abstract reasoning.

These two individuals were hardly unusual in their view of cognitive development, a view that has long been at the center of educational theory. For centuries, younger children have been described as qualitatively different in their manner of thinking, trapped in earlier stages of cognitive development. When children enter formal schooling at ages five or six, they have been typically characterized as conceptually hobbled because they have not yet escaped the limitations intrinsic to a particular stage. Thinking concretely has been an especially common way of describing such limitations and merits special attention first as we try to understand why it is so appealing but also so wrong.

The problems with consigning a child to exclusively concrete thought were actually apparent to some major figures said to be concrete to abstract shift proponents. In particular, John Dewey and Maria Montessori provide excellent case examples of how thoughtful, experienced observers rarely settled on the extreme view of concrete-only thought. We'll first look at their perspectives before turning to four specific reasons why the fallacy endures and why it is so mistaken.

Our examination of the case with concrete and abstract thought then leads to consideration of broader ways in which young children have been portrayed as having intrinsic cognitive limitations. Most of those views posit qualitatively different stages of cognitive development in which the limitations imposed by each stage sweep across all content domains. For example, some accounts put young schoolchildren in a stage where they cannot reason about transitive relations of length (e.g., if Adam is taller than Bill and Bill is taller than Chad, Adam must be taller than Chad). Stage views also usually maintain that the same children will fail in reasoning about transitive relations in all other content domains ranging from time to morality. A critique of these stage views leads to contemporary views showing how real conceptual change is possible. These new models of change do not characterize children as going through sequences of stages, each with distinct limits on children's minds.

In light of this newer emerging consensus about the nature of conceptual change and development, we can finally ask about the implications for wonder. Unfortunately, despite vaulting visions of national panels, on-the-ground school realities show that instruction often seems to be under the sway of classical accounts. Those accounts lead to instruction that works against wonder rather than supporting it.

To get a better sense of the long-standing intuitive appeal of young children as thinking concretely, we start by going back a couple of centuries.

The False Primacy of the Concrete

Over two hundred years ago, in 1810 in Switzerland, Johann Heinrich Pestalozzi, who had successively failed as a pastor, a politician, and a farmer, eventually became a hugely successful educational theorist and practitioner.[2] He described the need to initially teach materials "through the senses" at the concrete level and gradually move up to the abstract level. He linked this teaching progression to what he assumed was a developmental shift from concrete to abstract modes of thought.

Pestalozzi views were echoed over the years by some of the most dominant figures in psychology, philosophy, and education. Maria Montessori, founder of the enormously successful Montessori school movement, emphasized a concrete versus abstract contrast as central to her developmentally oriented curriculum, and today, Montessori schools throughout the world almost always refer to a concrete to abstract shift as critical to their teaching philosophy. John Dewey, widely described as a philosopher, a psychologist, and an education theorist, discussed the shift at length in a 1910 essay labeled "Concrete and Abstract Thinking." Other leading developmental psychologists of the twentieth century such as Lev Vygotsky, Jerome Bruner, and Jean Piaget described patterns of cognitive development that have been construed as descriptions of concrete to abstract shifts. Related discussions refer to holistic to analytic shifts and context-bound to context-free shifts. Across all these examples, a common theme emerges: young children (usually up to at least age six) can only think about things or events that they can directly perceive with their five senses (e.g., car, dog, jump, sneeze). All other concepts are abstract and beyond their cognitive reach.

Taken literally, a wholesale shift in children's minds from concrete to abstract thinking is a remarkable claim. It is as if their cognitive machinery was changing on a scale even grander than technology's shift from analog to digital storage. In just a few years, photochemical gradients were replaced by discrete records of pixels and physical traces of sound vibrations were replaced by vast sequences of 1s and 0s. Was the child's mind undergoing an even more massive shift? Do children's minds somehow completely change how they represent and think about everything? If such cognitive transformations were actually true, they would mandate radically different ways of interacting with children before and after the shift. They would justify treating young children as laboring under a tremendous deficit in which they can grasp only the shallowest renderings of reality. Where do such views come from, and why do they persevere in spite of ample experimental evidence that fundamentally contradicts them? How is this shift related to other views that limit young children by seeing them as passing through cognitive stages that sweep across the entirety of thought? An understanding of the appeal of these views will help us to see how blindly embracing them influences education and, ultimately, our opportunities to engage, through wonder, in exploration and discovery.

Distorting Dewey

Despite strong historical tradition of theories of a concrete to abstract shift, infants, toddlers and preschoolers clearly detect and deploy abstract concepts in their observations and actions on the world. We saw in chapter 2 how they can reason about agents in ways that imply a sense of highly general abstract patterns. Moreover, they can easily use their sense of these patterns to interpret completely novel situations. How can we reconcile this recent research with more classical views?

Dewey gives us clues as to how a reconciliation might be possible. In his essay on concrete and abstract thought, he explains the subtle nuances of the two forms of thought and how they interact in development. This is unfortunately a much-neglected part of his essay, which has been seen as endorsing the concrete to abstract shift. Instead, he focuses on the course of instruction of content rather than on developmental changes in children's minds. However, his comments on how concrete and abstract thought are always interlinked apply to both kinds of processes:

> The maxim enjoined upon teachers, "to proceed from the concrete to the abstract," is perhaps familiar rather than comprehended. Few who read and hear it gain a clear conception of the starting-point, the concrete; of the nature of the goal, the abstract; and of the exact nature of the path to be traversed in going from one to the other. At times the injunction is positively misunderstood, being taken to mean that education should advance from things to thought—as if any dealing with things in which thinking is not involved could possibly be educative.[3]

Dewey suggested that the concrete and the abstract must coexist at any stage in instruction. Good teaching cannot sensibly convey just the concrete without the abstract. The same coexistence perspective holds for cognitive development. In contrast to Dewey's emphasis on the coexistence of both forms of thought, most adults see a concrete to abstract developmental progression in which children can initially only think in terms of simple physical things that are in front of them.

Misconstruing Montessori

Disciples in developmental psychology and education can turn their leader's subtly formulated views into much starker simplified caricatures. When

Maria Montessori's writings and practices are considered more closely, they illustrate how describing her approach with a one-line concrete to abstract mantra undermines much of what makes her body of work so interesting. Just as Dewey had nuanced and insightful view of children's minds, many of Montessori's insights fit well with views of young children as active, innovative, and effective explorers of their environments.

Montessori grew up in Italy in the late nineteenth century. Intensive study of the sciences and mathematics as a teenager prepared her to attend medical school in Rome, a bold and unusual move for a woman in Italy at the time. She spent her early years as a doctor working with developmentally delayed children. This experience led to her broader ideas about cognitive development and education. In 1907, she opened her own school in Rome, where she implemented a new way of teaching that ultimately led to Montessori schools in many thousands of locations throughout the world.[4]

Montessori viewed children as progressing through a series of stages that she called *planes*. The first plane, lasting from birth to roughly six years, characterizes children as immersed in concrete things, especially in the first three years. Her background in biology and medicine may have led Montessori to see cognitive stagelike changes in quasi-biological terms. She discussed how the loss of baby teeth, which occurred at the beginning of the second plane at around age six, could be seen as a sign of that new stage.[5] She drew an analogy between caterpillar and butterfly stages of body development and different planes of cognitive development, arguing that teachers must "protect" the child's ability to fully learn in its current plane (becoming a bigger and better caterpillar) so as to allow development to fully complete its growth in that stage before transitioning to the next plane.

Montessori described children under three years as having a special sensitivity to patterns in the world. They learned about these patterns by "absorbing" them automatically.[6] Without thinking about an object or a pattern, the young child's mind was well prepared to unconsciously absorb immense amounts of information. Montessori saw this ability as fading away as children transitioned to the second plane. She may well have been noticing something like the infant statistical learning described in chapter 2. There is, however, no evidence of statistical learning declining after age six, although it may be more obscured by surges in other more noticeable cognitive skills.

In practice, the emphasis on concreteness led to Montessori programs in which children early on are immersed in hands-on activities with physical materials. Abstraction is avoided in the sense that teachers don't try to directly teach formal mathematical principles and operations, the laws of physics, or the foundations of biology. Instead, they embody some of those ideas in physical materials and the child's interactions with those materials. Montessori believed that materials should relate to real-life things and not playful pretend objects. For her, the real world was so rich and compelling that the primary goal of education should be providing ways of connecting children's thinking to its fascinating patterns.[7]

Montessori may have never really believed young children's internal representational capacities were solely restricted to concrete concepts; her emphasis was more on how early materials and settings should focus on physical objects and activities to foster learning. Such an educational practice doesn't in itself deny a young child the ability to acquire abstract categories and relations. Indeed, Montessori argued that much of mathematics was already present in the child's mind, waiting to be activated by the right sorts of physical materials. Scholars rarely discussed the specifics of internal mental representational formats until the cognitive revolution in the later decades of the twentieth century, a perspective buttressed by the emergence of computers as models of how information might be formatted and stored in a physical system. For that reason, it is not obvious what references to concrete thought meant in earlier years. Montessori did not confine children's conceptual abilities to storing mere concrete images and facts. She once said that "to teach details is to bring confusion, to establish a relationship between things is to bring much."[8] A focus on relations is often seen today as at the heart of abstract thinking. She repeatedly argued that education should not be attempts to implant knowledge but rather must be designed to amplify intrinsic interests and exploratory capacities of young children.[9] She stressed children's strong internal drives to understand the world as well as the need for them to have some degree of intrinsic prior knowledge to be able to learn.

Montessori described "awakening" an interest that is already present in the child. Teachers were to provide environments and materials that supported and facilitated those interests. Teachers should present children with materials that were "beautiful" and helped their imaginations take flight. The next chapter shows how Montessori's stress on intrinsic interest and her dislike of external rewards resonates strongly with current research on how to best motivate children in educational settings. When a child was immersed in a task, that child "becomes calm, radiantly happy, busy, forgetful of himself and, in consequence, indifferent prizes or material rewards."[10]

In many ways, Montessori was nurturing wonder. In her view, education was not the imparting of knowledge but finding ways to set off a "release of human potentialities."[11] She wanted to find ways to get children completely immersed in acts of self-determined "intelligently active" explorations of real-world phenomena. Good education was "the stimulation of seeds of interest already sown."[12] She stressed the great potential within each child just waiting to be released by the right situation and materials.

Many Montessori schools produce strong educational outcomes in their students.[13] This may occur because much of the curriculum focuses on helping children become active, engaged learners who become better and better at uncovering patterns in the world. It also encourages interactions with others who can provide more information to bring to bear on a problem. In doing so, it embraces communities of knowledge and learning from others. Unfortunately, when adults construe children as thinking concretely, they do often end up viewing young children as having a cognitive deficit that severely constrains what they can learn. Concreteness means imposing limits on young minds rather than Montessori's view of concreteness as providing levers for unleashing and inspiring much richer forms of thought.

Four Reasons Why Concrete Cognition Is So Compelling

Why is the concrete to abstract story so compelling? What makes it so resistant to extinction? Four interacting processes converge to promote and preserve this view of cognitive development.

The first process concerns related changes that do occur in language. Younger children do not typically acquire words referring to abstract relations, such as "randomness," "adaptation," and "inertia." Instead, their words mostly refer to simple physical objects and salient events. Yet, infants are sensitive to randomness and inertia and toddlers to adaptation, among many other abstract patterns and relations. Young children are ready to learn about science in surprisingly abstract, not concrete, ways. Indeed, sometimes the concrete details take much longer to master.[14] For example, preschoolers may believe there is tiny "stuff" inside animals that causes them to have their surface properties and to grow and act as they do, but not until medical school do most adults have clear concrete ideas of the body's internal parts and how they work together.

Without careful investigations of their nonverbal forms of thought, children's cognitive abilities may remain unnoticed by the informal observer in comparison to the specific words they happen to utter. Why then, do words tend to first apply to concrete entities? When learning words early on, it may be easier to disambiguate their meanings when you can point to physical examples.[15] Even if randomness is an easily accessible concept, it may be more difficult to unambiguously indicate a good example than it is to indicate a dog. Point to real-world objects or events such as a car or a fire. Now, try to point to instances of abstract concepts such as justice or arrogance. These seem easier to designate through language. But difficulties in designating instances do not mean that children are missing the concept. In fact, the ease of designating meaning through pointing is often challenging in the real world even for physical objects.[16]

Although the frequency of abstract words increases greatly from first words to language at age five, even the youngest speakers do utter some abstract words (e.g., *go, hi, no, give, uh-oh, pretty, all-gone, more, sleep, silly, booboo, none*). Abstract words are also often emotionally laden (e.g., *bad*).[17] Young children *can* represent some abstract concepts with words. Also, the first time a child says a word is rarely the first time that child had the corresponding concept. Words are typically mapped onto preexisting concepts rather than suddenly popping into existence or somehow created de novo out of spoken language. The philosopher Jerry Fodor made the provocative claim that virtually all concepts were innate and that vocabulary growth was simply learning to map words onto those preëxisting concepts.[18] We don't have to accept a strong version of that view to grant that we may have a concept in our minds for years before we attach a label to it. I recently learned the Indonesian word *mencolek,* which describes the prank of tapping someone from behind on the shoulder that is away from you so as to get them to look in the wrong direction. Every child over five probably knows exactly what this action is, but most English speakers have no word for it. English speakers have many other wordless concepts such as these two: *gigi,* a Filipino word meaning the urge to pinch or squeeze something that is overwhelmingly cute, and *tartle,* a Scottish word describing the mixture of panic and hesitation when one is just about to introduce a person whose name you suddenly can't remember. Many abstract concepts can be familiar to us and easily recognizable, but not encoded as single words. One can also generate new words for familiar concrete things at will. I might christen the debris left behind after raking a lawn full of leaves *ditsibits*, or a lone tree that sticks above all others on a tree line as an *arborstriver*. The absence of a word doesn't mean that something is abstract.

The second reason for viewing thought as initially concrete may arise from the cognitive effects of familiar, high-frequency words, such as *apple*, in comparison to the effects of unfamiliar low-frequency words such as *carburetor*. Younger children perform better on tasks using familiar items and high-frequency words than on tasks with low-frequency or nonsense words, but these benefits mostly arise because familiar words impose smaller memory loads. Loads lighten because preexisting mental links to other familiar concepts enable more efficient reasoning, not because young children can't have abstract concepts. Infants easily think about momentum, inertia, randomness, and intentionality without needing words for those concepts.

A third reason why developmental shifts in word meaning may also falsely imply a concrete to abstract shift occurs because of another actual shift. Children often learn the characteristic features of words like "uncle" (e.g., male friend of parents who is warm and caring) before they learn the defining features (e.g., brother of one of a child's parents).[19] But such shifts occur at different times for different semantic fields. They occur at around age seven for kinship terms but years earlier for moral transgression terms (steal, lie, hurt) and years later for cooking terms (boil, bake, fry). Even adults make the same kinds of shifts in novel domains (see figure 5.1).

Thus, this progression represents a common sequence at any age, not a developmental stage. Introductory college physics students sort problems into categories based on characteristic features while those more trained in physics sort the same problems into different categories based on deeper features.[20] All ages frequently adopt a strategy of initially relying on broad summary tabulations of many properties that co-occur with labeled instances before trying to determine which properties or underlying causal relations really matter. Importantly, the characteristic features are not always more concrete than the defining or causally central ones.

The fourth reason for inferring a concrete to abstract progression arises from equating the progression of instruction from novice to expert with a developmental progression, a point made by Dewey as well. Because instruction at any age often starts with concrete examples and then progresses

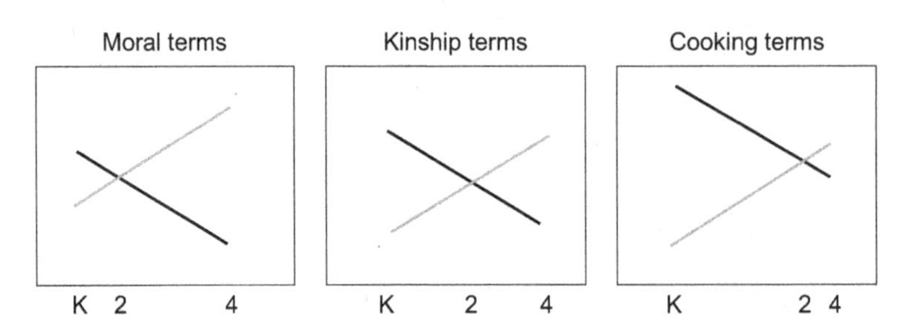

Figure 5.1
Characteristic to defining shifts in word meanings may sometimes be erroneously construed as indicating a general shift from a concrete stage of thought to an abstract stage. Children do shift from interpreting meanings in terms of large sets of characteristic features (darker lines) to meanings that focus on a few defining features (lighter lines). This happens at roughly the same time for most terms in a tightly related semantic field, such as kinship terms, but at very different times across domains. For words without clear definitions, analogous shifts occur from characteristic features to compact causal theories. (From Keil, F. C. (1989). *Concepts, kinds, and cognitive development.* MIT Press.)

to abstract principles, one might infer that cognitive development must unfold in the same way. In fact, adult instruction should not always start with concrete examples devoid of overarching goals or abstractly stated questions, but the common practice of starting with such examples can lead to faulty inferences about development. For example, for many familiar objects, such as a grandfather clock, you might start adult lessons by describing the internal mechanisms, assuming that adults know what grandfather clocks are and why people first created them. For a child, however, you might start with abstract descriptions of why we needed time of day displays and that grandfather clocks, because of their long cases, were the first relatively inexpensive clocks able to do this accurately. Only much later might you start to unpack the mechanism.

Evidence Against the Shift

Over the past few decades, researchers have consistently demonstrated the limitations of models based on monolithic concrete to abstract shifts. Early on, our lab group showed how young children were sensitive to abstract patterns distinguishing the insides of animals and machines long before they knew any details about what the insides looked like.[21] Similarly, in the introduction to her book *The Essential Child,*[22] Susan Gelman began with a discussion of how essentialism is contrary to a concrete to abstract shift. Children can see animals as having inner essences that are critical to explaining the presence of visible properties while at the same time having almost no concrete knowledge of what those essences are like or how they produce surface effects. We also now know that young children may agree on the relative mechanistic complexity of machines and animal parts while knowing little about what those parts look like. Abstract and concrete information processing complement each other in development. In addition, different cultural contexts can bias young children of the same age to engage in either abstract or concrete patterns of thought on the same tasks.[23] Current versions of Dewey's view argue that interactions between these two facets of thought must be considered throughout the teaching process at any age.[24]

Confusions about abstract and concrete thought matter because they can foster assumptions about children being "trapped" in concrete thought, which in turn lead to additional assumptions about instructional content, delivery, and assessment. Taken together, these assumptions can decrease student engagement with science. Instruction and assessment may ignore just those aspects of the world that children naturally find the most fascinating. Of course, abstract concepts that build on several others, such as *mortgage,* can take years to acquire. But such late acquisitions hardly discount the early emergence of other less culturally dependent abstract concepts.

Different Disconnects about What Develops

The concrete to abstract shift is just one of several traditional views of children's minds as going through a series of metamorphoses that, through a well-ordered sequence of steps, free them from profound limitations on how they think and what concepts they can entertain. Collectively, these views tend to posit stages of cognitive development. In addition to shifts from concrete to abstract, we find alleged shifts in stages from enactive to iconic to symbolic, illogical to logical, holistic to analytic, and context bound to context free, among others.

The most dominant cognitive developmental theorists of the twentieth century, such as Jean Piaget, Jerome Bruner, and Lev Vygotsky, all proposed such stagelike accounts. These metamorphoses of the mind involved complete reorganizations of how reality was represented. An activity such as *buying* might be first understood as a succession of visual images and then later be understood as a set of rules describing a transaction, rules that were purportedly outside the cognitive reach of young children.

Stage theorists almost invariably endorsed domain generality. Domain-general stages arise from central cognitive capacities manifested in all areas of thought. Stage changes were like swapping out a computer's CPU with a new model. The new CPU handles information in a completely different manner. Somehow, a "CPU" in the child's brain becomes transformed. A concrete child thinks concretely in folk biology, folk physics, and even morality, and when that child shifts to abstract thought in one domain, the same shift occurs in all other domains at the same time (with some slippage in synchrony being allowed by certain theories). But decades of study have failed to uncover such across-the-board strict representational deficits in development.

Several large-scale reviews have discussed mismatches between formal education and the informally developed capacities that children bring with them,[25] but the problem endures. Kindergartners are still characterized as trapped in earlier cognitive "stages." For example, children in the early school years seemed to neglect deeper relations at the heart of analogies and focus on surface matches.[26] Yet, more naturalistic stimuli presented in causally rich contexts can make analogies apparent to four-year-olds.[27] Similarly, when children have well-developed knowledge in both domains used in a metaphor, metaphors between those domains come tumbling out spontaneously. In a charming set of observations, the Russian poet and children's book author Kornei Chukovsky showed that preschoolers would coin completely novel metaphors such as "I'm barefoot all over," in reference to not having any clothes on.[28] Later researchers documented many more spontaneous early metaphors by toddlers, such as calling tires a "car's shoes."[29] Extensive knowledge in task-relevant areas can make a child look quite sophisticated in comparison with cases in which the child's knowledge is minimal. Differences in specific knowledge bases are often confused with alleged age differences in general cognitive abilities.

Younger children certainly can fail in tasks that older children pass, but their failures are typically for other reasons than their lacking a basic representational or computational capacity. Jean Piaget, the great twentieth-century Swiss psychologist, conducted studies in which children under roughly seven could not *conserve* various quantities across transformations. They supposedly lacked certain mental *operators* resembling logical rules. They were in *preoperational thought*, cursed with CPUs that could not employ certain computations and representations. Imagine that a five-year-old is presented with two rows of eight checkers spaced equally apart and is then asked if there are the same number of checkers in each row. The child says they are the same. Then, suppose the top row is spread out in full view of the child so that the spaces between checkers are 50 percent larger and the entire row is therefore 50 percent longer. If the child is then asked if one of the two rows has more or if they still had the same number, most five-year-olds say that the top row has more checkers. They failed to *conserve* quantity. In contrast, most eight-year-olds say that both rows obviously still had the same number because the experimenter hadn't changed anything important.

Alternative Explanations for Global Stages

Piaget explained these dramatic shifts in conversation ability in terms of the younger children missing certain mental *operators*. Operators were cognitive reasoning tools that enabled children to think in terms of conceptual relations such as identity, compensation, and reversibility. But hundreds of follow-up studies revealed that younger children were not failing because of missing mental operators, but because other aspects of the task unrelated to Piaget's theory were misleading them. These factors included limited attentional skills (e.g., focusing too much on one dimension of the task), lower memory capacities (e.g., answering correctly only when the number of checkers is small), and different social pragmatic inferences (e.g., because you intentionally changed something, you must want me to change my answer from same to different). When later experimenters controlled for these other factors, much younger children could conserve. Conversely, increasing the load of such factors can cause older children and even adults to fail.[30] These developmental changes are now usually interpreted as arising from gradually improving abilities in memory and attention as opposed to abrupt stage transitions in how concepts themselves are formatted. When more sudden changes do happen, they occur as conceptual insights in well-bounded domains.

What causes developmental changes in memory and attention? Many researchers point to growth of *executive functioning*. Executive functioning is the interrelated set of skills revolving around the deployment and regulation of attention in tasks with an explicit goal. Such goals include memorizing a list of words, focusing on a target and ignoring distractors, and resisting a first impulse to make a response. Those with high levels of executive functioning tend to be more cognitively flexible (e.g., taking different points of view on a problem), have larger working memories (e.g., keeping several bits of information in mind at the same time while manipulating them, as in mental math), and have greater inhibitory control (e.g., blocking a common but wrong response in favor of a less common but correct one or resisting the temptation to attend to a distractor).[31]

Executive functioning skills involve the prefrontal cortex of the brain, an area that develops relatively late compared with other brain regions. Most researchers believe that executive functioning, while subject to maturational constraints, is also malleable in the preschool and early school years through practice on executive functioning tasks and limiting exposure to excessive stress. Finally, lower socioeconomic status (SES) children tend to have more limited executive functioning skills, presumably because of less opportunity to practice them.[32] If a child is in a chaotic day care setting with far too many children per caregiver and few material resources, that child simply may not have the opportunity for extended guided interactions with another or even for quiet engagement with an interesting toy or educational game.

In short, developing attentional and memory skills, many of which are related to improvements in executive functioning, can explain away many apparent stagelike phenomena. When these are combined with increasingly sophisticated pragmatic interpretations of what adults want in various tasks, global stage theories fall apart. Children's cognitive abilities clearly do change from infancy onward, sometimes in ways that can be quite dramatic when one is looking just at their surface behavior. Understanding what lies beneath these surface changes is more difficult to discern, and people can become captivated by ideas such as across-the-board stages of cognitive development and shifts from the concrete to the abstract. Such mistaken views may not matter much when merely speculating about young minds, but they can be destructive when they lead to specific pedagogical practices. Before considering those problems and how we might address them, we need a brief sense of more current accounts of how knowledge and conceptual understanding do grow during the preschool and early school years.

Contemporary Accounts of Knowledge Growth and Conceptual Change

From infancy onward, children's knowledge grows in both amount and structural complexity. Most current researchers agree on certain general features of knowledge growth. These features have emerged over the last few decades as studies probed ever more closely how knowledge increased in both depth and breadth.

Patterns of Knowledge Growth

The simplest form of knowledge growth is the mere accretion of bits of information, such as learning an ever-larger list of Pokémon character names (898, at the time of this writing) without knowing anything about them. Some younger siblings of Pokémon fans may attempt to accumulate such mindless lists.

Whenever possible, we go beyond simple lists to seek structure to relate new items to each other or to connect with our existing knowledge. We relate fictional characters together through their interactions in stories or through their typical locations. Or we link them to real-life individuals or characters in another genre. We build up larger interrelated knowledge structures because they provide well-known cognitive economies in which each element serves as a reminder of another and which often form meaningful "chunks" in memory. The classic demonstrations of chunking involved chess masters who stored common sequences of moves (e.g., variations of the Sicilian defense) as one chunk.[33]

Many kinds of structure can support knowledge retention and growth. We map items to be remembered onto familiar real-world spaces such as different locations in our homes, a method used by Cicero to remember his long orations in the Roman senate. We embed them into rhyming verse or alliterative phrases, thereby using structure from language. We put them into informative hierarchies, as we often do for animals and many artifacts. We link them causally. We can do many of these things at the same time, such as using converging support from spatial, hierarchical, and causal relations, as occurs in mechanistic representations.

Rich knowledge structures can enhance memory and attention in ways that help transcend the limitations of weaker executive functioning skills, those abilities to deploy and regulate attention. A dramatic improvement in memory was shown by the psychologist Micki Chi with five-year-old chess experts (every large city has at least a few). These tiny chess experts could easily remember far more pieces on a chessboard representing a game in progress than could adult novices. The same memory skill held for young dinosaur experts.[34] Child experts were able to embed new memory tasks into their preexisting knowledge structures and use that structure to enhance recall. Success, however, was closely tied to the specific domain involved. The child chess experts had far better recall of chess piece positions than did adult novices, but they would do far worse than the same adult novices when asked to recall positions of checkers on a checkerboard (assuming they were not experts in checkers as well).

Knowledge also grows in larger domains, such as those covered by core knowledge systems. Children's knowledge of physical mechanics, folk psychology, and folk biology expands greatly from the early preschool years onward. Animals may become organized into classes at levels that go both further up and further below the basic levels of dog, cat, cow, and horse. A child's concept of dog may differentiate into knowledge of different breeds and subbreeds. At the same time dogs may be understood as members of a larger class of canines, which in turn are understood as types of mammals. At each level, children build interconnected elements of knowledge that help properties hang together. They might learn how dogs differ from cats in terms of social interactions with humans. They might learn surprising commonalities among all mammals despite massive superficial dissimilarities such as those between a whale and a mouse. Conceptual change takes on a life of its own that is unique to the domain involved. The causal regularities in a domain such as biology exert their own special influences on how knowledge structures shift over time.

The development of knowledge structures is subject to soft and hard constraints. Soft constraints include the weight of evidence, such as the acquired knowledge that all encountered mammals have easily observable limbs. This evidence can make it initially difficult to think of a dolphin as a mammal. Hard constraints are often related to core domains. Physical solids cannot interpenetrate each other or act on each other at a distance. We have greater difficulties understanding violations of such constraints, such as telekinesis. Executive functioning skills may be critical to overriding soft, and even hard, constraints when needed.

Implications for Models of Conceptual Change

When researchers discuss knowledge growth, they often refer to *conceptual change*. Conceptual change has several distinct meanings. It can mean gradual differentiation of a structure such as adding lower and lower branches to a hierarchical tree about animals. It can be a generalization, such as assuming that a new mechanism explaining digestion in cats probably applies to all animals. It can mean a sudden insight that a well-worked-out system of knowledge is relevant to a new domain. Imagine realizing that coral is a kind of animal and not a rock and suddenly transferring an entire set of new biological causal relations concerning animals to coral. Conceptual change can also mean shifting from using a large set of typical features to describe a category to focusing on a few critical features. This can occur as a characteristic to defining feature shift in the understanding of *uncle*.[35] In this example, an entire *semantic field*, namely all kinship terms (uncle, aunt, cousin, grandfather, etc.), shifts at roughly the same time.

Finally, conceptual change can be a conceptual revolution that causes radical restructuring of domains. The historian of science Thomas Kuhn famously conveyed this idea in his book *The Structure of Scientific Revolutions*.[36] He described transformative shifts in understanding, such as realizing that fire did not consist of a special stuff called phlogiston that was released when combustible substances were heated. Other examples described include the Copernican revolution of reconstruing the heavens as not revolving around the earth, the Darwinian revolution in which species of animals were understood as arising from the process of natural selection and not from the designs of a deity, and the revolution overturning the view that light must travel through a physical medium called *luminiferous aether* to conceptualizing light as electromagnetic radiation with properties of both waves and particles.

Do such conceptual revolutions occur in cognitive development? Revolutions do seem to occur, but other forms of conceptual change can often masquerade as illusory revolutions.[37] For example, a shift in understanding of coral might look like a revolution until one realizes that it is instead the realization of the relevance of another well-established set of beliefs. When genuinely new forms of knowledge or understanding do appear, they often occur when two domains are brought together and create a new emergent form of understanding nothing like those two inputs. Such shifts may occur when a child uses expanding linguistic skills to combine components of arithmetic understanding to create a new understanding of the concept of a natural number and when infants combine simpler elements involving agents and "social beings" to create the new idea of "social agent" at around 12 months of age.[38]

All these cases of change, whether modest and incremental or dramatic and revolutionary, do not easily map onto any idea of a concrete to abstract shift. Abstract and concrete properties and relations are usually intertwined in various ways throughout all kinds of conceptual change. A child who learns that whales are mammals is not just learning new concrete information or new abstract information. The child is learning how interactions of both apply to a new class of entities. Sometimes concrete features may predominate early on, but, other times, the child may start with abstract ideas about a kind and then spend a lifetime filling in concrete details.[39]

Most current theories of conceptual change would therefore accommodate several distinct types: differentiation, generalization, sudden insights such as through transfer of causal relational patterns across domains, shifts in feature valuations, and conceptual revolutions precipitated by recognition of growing internal conflicts. It is, moreover, not always obvious at first impression to tell which kind of change is actually occurring. In addition, several types of change can occur at the same time. Informed careful observations, and often extended interactions with a child over time, are usually needed to know what kinds of conceptual change are occurring. We can however, reduce pedagogical harm by rejecting classic stage accounts no matter how intuitively appealing they are and no matter how easily they can be encapsulated in a brief statement.

Classroom Consequences of Deficit Mirages

The well-documented cognitive activities of infancy and early childhood provide powerful skills for learning science in elementary school and beyond. Mistaken views of children's minds can undermine those early abilities. When children enter the classroom, they often encounter educational approaches clashing with how they have been learning from early infancy.

Assumptions of domain-general stages lead to views of children as having widespread limitations imposed on them by their stage. This leads to "deficit" views of children's thinking with education having the task of addressing those deficits.[40] The deficits might be a list of specific misconceptions, such as thinking the earth is flat. They are presumed indicators of a qualitatively different immature mode of thought, such as the concrete level. When parents, teachers, children's book authors, and science museum staff focus on what children can't do rather than on what they can do, they will tend to see science instruction for young children as consisting of concrete simple facts and perceptually exciting demonstrations. Following that lead, many assessments focus on recall of facts and surface relations.

Teachable Science and Statements of Fact

An emphasis on the concrete (or other related stage constructs) can lead to equating the teaching of science with imparting explicit statements of facts about physical things: the sun is hot and very big, spiders have eight legs. Teaching this way reduces learning opportunities for children, and those opportunities are far more than just the implicit learning of associations and correlations. They also involve learning about abstract causal relations and how to connect them together to form models of the world. Thus, a child, after systematically playing with a novel toy for several hours, might develop a model that it has a battery-driven motor and that the battery gradually loses its energy, which makes the toy slow down. This, however, does *not* mean one should just omit didactic instruction and let children explore and discover on their own. The psychologist David Klahr and others have shown clear benefits not just of guided exploration but also of direct delivery of facts and causal relations, provided they are embedded in appropriate goals and learning contexts.[41] For example, schoolchildren make more thoughtful and creative science fair posters when guided by some degree of direct instruction. Fully engaged partners in exploration and discovery can greatly accelerate learning, especially when those partners know how to gently guide the child toward learning the most important and central properties of the object of interest. This is another version of Margaret Donaldson's warning (see chapter 1) that most schools are instructionally mismatched with children's minds.

When instruction focuses mostly on learning concrete and easily statable facts, it can contaminate assessments. Far too often, testing simply consists of regurgitation of those same facts. The most impressive and enduring forms of learning may be acquiring abstract causal patterns and implicit senses of the kinds of mechanistic systems that are in play. If children are only tested on their recall of factual details, they may be discouraged when they fail to recall everything. But they might be encouraged if they are also tested for their newly acquired intuitions of complexity, for their judgments of the relative plausibility of different explanations, and for their abilities to relearn old content or to acquire related new content.[42] Asking children for names and colors of engine parts may cause frustrating failures. Asking them which of two engine configurations would work better can reveal substantial learning.

If we assess children on those aspects of cognition where their learning is often weakest (spitting out facts) and, at the same time, neglect those aspects where their learning is more intuitive and stronger, we are setting them up to fail. After all, even preverbal infants can learn about abstract relations and use that knowledge to reason about novel events. Young children certainly cannot always learn what older children can learn. When retention of concrete details is taken as the essence of what is learned, factoids become the central focus of tests and can demotivate those learners who previously most enjoyed learning science. Assessments must not be construed just as educational quality control checks; they are deeply embedded in children's educational experience and can heavily influence their learning goals and, ultimately, their enjoyment of learning and their tendency to continue learning on a lifelong basis.

Lost in Translation—When Good Educational Intentions Aren't Enough

Deficit views of children have subtle, but corrosive, influences on even the best intentioned content. As well-conceived curricular guidelines are translated into actual practice by thousands of individuals at various levels of the educational system, the final results can be disappointing to everyone. Consider some of the materials commonly used in kindergarten and the first grade. The Next Generation Science Standards (NGSS) were released in April 2013. The NGGS publications contain inspirational statements that build on the studies described earlier. I and many other colleagues served on national committees that contributed to the reports leading up to the creation of the standards.[43] The NGSS stresses continuity of core disciplinary ideas over the years, such as, "Plants, animals, and their surroundings make up a system." They propose elaboration and deepening of those ideas in *learning progressions*. The proposed standards used soaring language stressing causal and mechanistic thought and use of that knowledge to build understanding both within and across disciplines.[44] The core disciplinary ideas cover a broad spectrum: Physical Science, Life Science, Earth and Space Science, and Engineering. (It wasn't clear why Earth and Space Science was listed as a foundational discipline in the same sense as Physics, Biology, and Engineering.)

A second commendable emphasis was on "cross cutting concepts" such as the relations between cause and effect, which are relevant to all disciplines. Finally, the NGSS materials emphasized learning scientific practices such as how to collect and evaluate evidence, formulate hypotheses, and test them. The importance of abstraction is stressed even in the early school years, but the idea is not developed and is not evident in most of the actual teaching materials. You might expect a great deal of attention to the cognitive endowments that children do bring to the classroom such as their knowledge of biological and physical phenomena and their grasp of the distinct properties of artifacts. You might also expect extensive use of the skills that preschoolers display in seeking out and evaluating sources, information, and the structure of explanations. You would surely want to take advantage of their abilities to navigate divisions of cognitive labor and think about causal patterns. All of these might then be used as a launchpad for further instruction. But despite clear acknowledgment of these early skills in the NGSS publications, they do not figure much in the curriculum.

The NGSS website fully acknowledges the challenges of creating good curricula and assessments but provides few example lessons. It offers equally few examples of assessments. Instead it provides guidelines and principles for designing materials and tests. This approach is laudable if it fosters local creativity and relevance, but it also runs the risk of having instruction be distorted by mistaken views of cognitive development and by vendors with agendas beyond educational excellence. Implicit and explicit assumptions about stages, concrete thinking, and deficits are prevalent. Even the NGSS's own guidelines for disciplinary core ideas in kindergarten seem to grossly underestimate what children bring to the classroom. For example, one of the core ideas to teach kindergartners is stated in the NGSS as "When objects touch or collide, they push on one another and can change motion." What is the point of this lesson? We know from many studies that preverbal infants in both their violated expectations and actions seem to have clearly mastered this knowledge and far more. During the preschool years we have seen that they can even envision counterfactual trajectories.

Is the point of teaching this idea to kindergartners getting them to explicitly state it in ways that foster the idea of forces? Perhaps, but this is not made clear to teachers. Science materials for the first several years of school often seem to be repeating beliefs about the world commonly held by preschoolers such as that pushes and pulls have different effects, that weather forecasts can tell us if bad weather is coming, that we need flashlights to see most things in the dark, and that larger objects can be built up out of smaller pieces. All of these might be unpacked in interesting ways, but that rarely occurs.

Assessments are equally discouraging as they are often focused on recall of facts or definitions. The Virginia State Department of Education is, unlike their counterparts in most other states, willing to post state science exams from prior years, exams that were meant to resonate with NGSS goals. Here are a few example items from a third-grade test:

> Which of these BEST describes the texture of the metal spoon?
>
> A—Small B—Smooth C—Flexible D—Shiny
>
> When notebook paper is folded to make an airplane, what physical property of the paper changes?
>
> A—Mass B—Weight C—Shape D—Smell
>
> What does a weather vane show?
>
> A—Wind direction B—Air temperature C—Cloud cover D—Air pressure

All three items are nothing more than science vocabulary tests for the terse meanings of texture, mass, and wind vane.

Other items are mystifyingly obvious from life experience:

> Which of these BEST describes the way a feather feels?
>
> A—Colorful B—Soft C—Low D—Bright

This last question is based on the teaching goal of telling children they can gather different kinds of information from each of the five senses. But does this item really assess an important new idea arising from classroom instruction? Third graders, and even middle schoolers, are often tested for memorization of isolated facts, definitions, or simple noncausal relations, such as distance differences. When this happens, a message is conveyed about what you should take away from your classes.

Much of early instruction consists of asking children to sort by different properties such as shape, size, or color and show the ability to use simple tools such as rulers and scales. Counting is also common. Causal explanations are rare. Even though the instructional goals are often posed in lofty terms such as "How does the structure of matter affect the properties and uses of materials?" the actual exercises don't rise to nearly the same level.

The guidelines developed by school districts to frame the NGSS and go beyond it often reveal assumptions about stagelike limitations. The school district in my own town in Connecticut describes the following sequence, which are also listed by many other districts.[45]

Grades K–2: "Development of wonder about the natural world and the ability to observe, describe and apply basic process skills"

Grades 3–5: "Development of descriptions of basic natural phenomena and the ability to perform simple experiments and record accurate data"

Grades 6–8: "Development of basic explanations for natural phenomena, and the ability to ask good questions and apply experimental procedures to collect and analyze data"

Grades 9–10: "Development of interest in global issues and the ability to collect, analyze and use data to explore and explain related science concepts"

Do we really have to wait till grade 6 or later to guide children on how to seek out explanations and ask good questions? We know that preschoolers are eager to hear good explanations. Moreover, they have strong preferences for certain kinds of explanations. They also often ask the right kinds of how and why questions and expect good-quality answers in return. They can certainly offer descriptions of many natural phenomena and do perform simple versions of experiments. It is less obvious how interested they are in global issues, but they certainly are capable of appreciating such things as pollution and rising sea levels. Finally, we don't need to instill or develop wonder in the first few years. We should be more concerned with protecting and nurturing the wonder that is already in abundance. Instead of suggesting a sequence that seems to reflect stagelike assumptions of maturational readiness, it makes more sense to see explanations, descriptions, interest, and wonder as endowments that children bring to the classroom. Education should help them build on these impressive abilities and show they can interact in increasingly complex ways. Education does not have to implant these skills in helpless kindergartners. All four are basic to how children try to make sense of the world from an early age.

Appreciating What Children Bring to School

When adults no longer view children's minds being imprisoned in broad stages of cognitive development, they start thinking about other ways children could learn and how that might best occur. They move away from focusing just on explicit forms of knowledge and de-emphasize notions such as maturational readiness. They move away from thinking of young children as bundles of "misconceptions" that must be erased and supplanted with correct conceptions and move toward building on children's already present abilities to make sense of things around them.

What should early science instruction look like instead? The curriculum could still embody much of the spirit of the NGSS broad goals, but it would be a more effective platform for future learning if it acknowledged more of what children know and can do and allowed them to experience and even embrace phenomena in all their complexity. It should also employ assessments of the most valuable and useful insights that students acquire from their instruction. A wide array of instructional strategies could be employed, many of which center around the idea of providing children with more elaborated explanations of phenomena.

We should help children sharpen their own abilities to see how the natural and physical worlds sort themselves into meaningful domains because of common underlying causal patterns. Many kindergartners might not know that whales and dolphins are not fish but actually mammals. The research literature suggests how to explain insight to them. Provide them with information about the underlying processes and causal pathways in common to mammals and remind them why that information is especially powerful. Similar exercises could be done for shifting coral to the animal category and for lower level re-sortings such as moving legless lizards out of the snake category. Always accompany information with explanations that link properties together.

We should never try to hide complexity. Allow children to marvel at the complexity inherent in nature much as scientists did centuries ago when a whole new area of discovery opened up. Lift up the hood and allow them to see the machinery or mechanisms inside. At the same time, show them how one needn't know all the mechanisms in full detail and explain that in many cases no one person, and often no group of people, does. Instead, they should be taught how it is possible to achieve meaningful understanding at coarser levels and indeed to appreciate the idea of levels and layers of explanations. This reassures them of the feasibility of their learning goals and also of an almost endless future of exciting learning and discovery.

At the same time as we reveal all the complexity and its layered intricate structure, we might also try to demonstrate the coexistence of overarching elegant simple unifying principles such as conservation of energy, a common code for all living organisms, and the value of stable subassemblies with distinct functions in complex systems.[46] In doing so, we can help students better understand the value of idealizations and models. Science can be at its more glorious when we appreciate a thing's immense complexity while also realizing the elegant overarching principles that shape it. For example, in explaining all the details of how an internal combustion engine works, we might also point out that they were created as a way of harnessing the enormous energy contained in the bonds of certain chemical compounds (e.g., gasoline) by transforming uncontrolled explosions into easily regulated rotating shafts.

Assessments of learning should not overemphasize definitions and fact retrieval. Such things are useful to the extent that they are connected to larger sets of causal beliefs and metadata. Children need to see how abstractions and concise gists can enable them to pick out good versus bad experts, to know what is relevant, to relearn old content, to more easily learn related new content, to ask more probing questions so as to expand understanding, and to keep refueling that glorious feeling of wonder. Some forms of instruction can fit well with such an orientation. For example, instruction might employ ways of encountering distinctions that occur within contrasting cases. As proposed by the education researcher Dan Schwartz,[47] contrasting cases present two different kinds of things that initially seem to be the same thing.

Using contrasting instances, a teacher might present a bottle-nosed dolphin and a shark of the same size. By asking students to look at both cases closely, they may start to see what differences are present and why they matter. They might note from videos that they swim in different ways and that the dolphin's pattern of swimming fits well with activities around its blowhole. With a little guided exploration, they might come to realize that whole clusters of properties are meaningfully related in dolphins in ways that are very different from how other clusters are related in sharks. This kind of instructional technique can naturally help build more abstract knowledge and fits well with the best kinds of assessments.[48] The contrasts have to be thoughtfully designed and may require some subtle supporting pointers, but, at their best, they may trigger new well-articulated wonders about each topic. A feature-rich contrast between superficially similar cases may thus foster questions about what kinds of deeper contrasting mechanisms are involved. Such approaches can even enhance metacognitive awareness as they enable learners to evaluate how they need to reflect and go beyond their first impressions.[49]

Much of science education today also focuses on basics of experimental design, data interpretation, and the nature of science.[50] Such approaches may be largely orthogonal to wonder. Preschoolers don't need to know much about how to do science. Yet, growing wonder may promote these skills anyway. As why and how inquiries probe deep and deeper, one will inevitably start thinking more directly about evidence and methods. Similarly, metacognitive awareness will increase.[51] Less time may be needed for explicit instruction on how to do science and the nature of science if a child has developed a deep and well-justified grasp of mechanisms. In my years as an undergraduate at MIT, no science or engineering majors that I knew took courses on the nature of science or general methods. They had not covered such topics in high school either. Having been entranced with science from an early age, however, they had assimilated much of what was needed. For example, a child notices that an alert goes off on a new security detector every time someone knocks at the front door. He might wonder if it is noise of the knocking or the visual movement that triggered the alarm. He might experiment by standing at the door pretending to be knocking but not making any noise and contrasting that case with playing back a recording of the knocking noise without anyone standing at the door. If only the movement matters, he might explore which aspects of the movement make a difference. If birds start to gather around a new bird feeder in a new location full of seed, he might vary the shape, color, and taste of the feed to see if one feature was most important.

These informal experiments might not always entail perfect de-confounding of variables, but with a little support from parents and others and a drive to understand mechanism, they can emerge quite naturally and become more and more sophisticated. A great many children at some point experiment in groups with different paper glider designs, trying to create one that will fly the longest straight course when thrown from the same height. With a little attention to different variables such as nose weight, wingspan, wing shape, bottom vane, and wing slant, a group of inquisitive children will often start to more carefully control each variable independently as well as matching initial launch conditions. Sometimes, they may need a little prompt from a parent or other interested adult who points out two things that had been changed at the same time and need to be disentangled. Over time, and with some support from others, asking such questions about relevant variables became a natural part of thinking about both biology and devices. Admittedly, this isn't always easy. The number of possible variables can be overwhelming, and they might have complex interactions with each other. In such instances, helpful adults can provide invaluable advice in how to start teasing things apart.

Beyond the Schools

All of these ways of learning extend far beyond the classroom. Science museums are terrific examples of potentially informative settings. Throughout the world, science museums have beautiful eye-catching exhibits and often promote spectacular events in which things go poof in flames and smoke; chemical mixtures change rapidly in size, color, and texture; and things levitate. I've been to dozens of science museums as a visitor, have been to many others as a researcher, and helped build another from the ground up. I have greatly enjoyed the eye-popping shows as well. Unfortunately, some science demonstrations only offer lip service to the science behind the display and present it in a manner similar to those speeded up sotto voce cautions at the end of drug commercials, obligatory but rushed through and easily ignored so as not to detract from the big show. In one common display, demonstrators' and audience participants' hairs stand on end while touching a Van de Graaff generator, but often no coherent explanations are provided as to why. The dangers of such shows arise when the science part is subordinate to the "wow" component and is largely forgotten. Such explanation-free demonstrations can mislead audiences into thinking science is all fun and games and akin to a spectator sport. This may set up some people to fail if they start to study science with that preconception. Moreover, such demonstrations can condescend and patronize rather than truly engage and uplift the public through inspired wondering.[52]

In contrast, when done well, science museum exhibits serve to "engage, educate and empower," which is the goal of Ithaca, New York's, Sciencenter. This perspective was inspired by the Sciencenter's founders Debbie Levin and Ilma Levin who spent fifteen years developing the approach through their

Figure 5.2
Debbie Levin and Ilma Levine, founders of the Sciencenter in Ithaca, New York, at its initial construction in 1992. Levin and Levine made sure that exciting demonstrations never eclipsed or displaced the fascinating science behind the display.
Source: Sciencenter, http://www.sciencenter.org/perch/resources/209318-sc-casestatement2015-pages-cropped.pdf.

volunteer science programs in Ithaca's elementary schools (figure 5.2). They learned how to combine engaging demonstrations with thought-provoking and wonder-encouraging exploration. Their goal: "to inspire curiosity and passion for science in children."

Including a fun component in science demonstration is fine as long as it doesn't displace the science. I am reminded of how "fun" additions to classical music can occur without diminishing the music. At Yale each year a much beloved Halloween show is presented by the gifted members of the Yale Symphony Orchestra. The orchestra members get dressed up in outlandish costumes and play beautiful classical and popular music as the "soundtrack" for a silent movie that parodies Yale or current affairs. It is always a sold-out event and includes inspirational classical performances. No one sees this as pandering to baser instincts, but rather as a playful way to remind the audience of the incredibly talented musicians who are part of our community. For similar reasons, science presentations that use dramatic displays to entice the audience shouldn't always be seen in a negative light.

In middle school I attended a lecture in which the speaker started with graphic videos of Venus flytraps closing on their live prey, often with details of them struggling in vain to get free of the entombing "bars" at the end of the leaf. Combined with dramatic music, this opening got our attention. But what made the lecture so entrancing and memorable almost sixty years later was what came next. The speaker asked how a plant could do that. How did it "know" the fly or bug was in position to close the trap? How did it move the trap? Did it have a brain? Did it have muscles? Through extensive back-and-forth banter with a raptly attentive group of children, he pulled back the curtain and revealed some of the complexity of underlying mechanism. He even raised the question of how this system could evolve. The elaborated story is amazing and is still being unpacked as shown by a flurry of papers about newly discovered molecular pathways involved in prey detection and trap closing.[53]

The flytrap demonstration worked because it never tried to hide the complexity of underlying mechanism. It reveled in it while also describing some elegant principles that helped make sense of the complexity. I have been fascinated by carnivorous plants ever since. I don't underestimate the skills needed to take content of original journal articles and make them both sensible and entertaining to the public. Consider some of the language in the abstract of a 2016 paper on flytraps in *Genome Research*:[54]

> The transcriptomic landscape of the Dionaea trap is dramatically shifted toward signal transduction and nutrient transport upon insect feeding, with touch hormone signaling and protein secretion prevailing. At the same time, a massive induction of general defense responses is accompanied by the repression of cell death-related genes/processes. We hypothesize that the carnivory syndrome of Dionaea evolved by exaptation of ancient defense pathways, replacing cell death with nutrient acquisition.

Although some biologists might be drawn in by the term "transcriptomic," I'm guessing most people do not find the language inviting. Gifted science writers are somehow able to take discoveries from highly technical journals and turn them into accessible prose that reveals the full beauty and complexity of the system and its mechanistic underpinnings, but such talented interpreters are rare. It is fine to start with flashy spectacular introductions, but only when the curtain is also pulled back to showcase the marvelous machinery of nature. Those revelations nurture old wonders and promote new ones. If that flytrap speaker in 1962 had just shown a gory clip and nothing else, I'm quite sure the excitement would have quickly faded and the entire demonstration would have disappeared from memory.

Most science museums, or TV shows, or public speakers don't intend to condescend in the sense of deliberately making things more simplistic than is appropriate for their young audiences. Instead, the problem usually arises from pervasive views of children as concrete thinkers, trapped in earlier stages. With such discredited views, adults try to speak to children in "their language." Unfortunately, all too often those mistaken beliefs about children's minds result in shows that are little more than smoke and mirrors.

How we explain matters. The best explainers either know a great deal about the topic themselves or at least know who knows. They have learned what kinds of gists support, rather than thwart, paths to deeper understanding. They are happy to say they don't know how to answer a question. They then describe who can provide the answer or, with an air of excitement and suspense, reveal that no one yet knows. Modeling this form of intellectual humility is in the best tradition of how scientists themselves talk to each other.

Obviously, children are not just like adults. Kindergartners wouldn't thrive in college courses if only we had more positive views of them. Families of differential equations is not a good topic for kindergarten. Curricular appropriateness is related to how knowledge builds on other earlier learned components, just as a roof cannot be built without first building the supporting walls. You can't learn about differential equations without knowing basic calculus, which in turn requires algebra.

Adequate working memory capacity (roughly how much information one can consciously keep in mind at one time) is also important. For example, learning how to construct proofs in formal logic requires a well-developed ability to keep several relations in mind at once, an ability that sadly peaks in the late twenties. This is why many older adults who studied logic and advanced mathematics in college will sense an ever-smaller mental workspace on which they are able to manipulate several formal relations at once. As a consequence, they may feel increasing difficulties in solving novel complex proofs. They can often compensate in clever ways, but the challenges faced by older scientists help us understand related challenges faced by the young child.

Some Early STEM Learning Guidelines

We have traversed a lot of territory, ending with a somewhat dispiriting description of what often goes wrong in schools and other informal settings concerned with science education. Wonder is neglected, and children may come to dislike and avoid situations that should inspire its growth and attendant pleasures because those situations often do the opposite. A number of positive alternatives have been mentioned, but at this point, three suggested guidelines arising from our discussion might be useful for anyone interested in improving STEM learning.

First, focus on what children can do, not what they cannot do. They have marvelous cognitive abilities that should resonate, not clash, with curriculum and teaching practices. If you or someone else suggests that younger children are simply unable to comprehend a particular kind of concept or knowledge, think again about all the ways that they might be having difficulties that have nothing to do with being in an earlier stage. Rather than thinking of them as a different cognitive species, it may be helpful to imagine for a moment that they are more mature humans from an utterly different culture. You wouldn't cognitively condescend to such people, but you would focus more on how they might be interpreting the learning situation. It is not a fully accurate analogy, but it may help you rethink what is going on.

Second, do not put yourself either in the role of *broadcaster* of information who measures success by measuring how much is recalled by students or in the role of *cheerleader* who warmly encourages children to do whatever they want and to discover things on their own. Either extreme is far worse that seeing yourself and each child as partners on shared quests to uncover underlying interpretable causal patterns. This doesn't mean strictly equal partners in all respects, but it acknowledges that both members of the dyad have much to learn and that each can help contribute to the process of exploration and discovery in a collaborative manner. It is difficult to do this effectively in a large classroom with every one of twenty or more children at the same time. A few especially gifted teachers seem to find ways to set up the process of discovery at several levels so that every child can engage and gain insights throughout a class session, but no one should underestimate the skill and effort required.

Finally, remember how much children in the home love to ask how and why questions and get meaningful answers. Keep track of how many such questions spontaneously arise in the classroom and work hard to come closer to the rates observed at home in the late preschool years. Ask questions yourself not of the students but as an act of wonder and then jointly share with your students ways to find an answer, all the while encouraging them to refine the question or ask new ones as the first gets partially answered. In doing so, you will naturally be focusing on underlying causal patterns and clusters and not on mere lists of concrete facts.

▪ ▪ ▪

Appearances can be deceiving. It is easy to see the appeals of views endorsing concrete to abstract shifts, domain-general stages, and deficit models. All of these views, however, are oversimplified distortions. Much of what develops involves increasing executive functioning and knowledge growth, usually in interactive mutually supportive cycles. Knowledge differentiation and conceptual change are most apparent in bounded domains and not as wholesale replacements of the brain's representational and computational architecture. Mistaken views heavily influence how adults think science should be taught and presented to children. When this happens, the problem is not just ineffective or misleading instruction but, even worse, that the mis-framing of goals and assessments toxically undermines motivation and stifles future conceptual growth.

When we view young children as laboring under profound cognitive limitations that make some kinds of thought impossible for them, we hinder what they want to do most. Their natural drives to explore, discover, and learn may be crushed. Imagine if the only information you could receive and discuss with others consisted of facts about perceivable objects and events. It would soon drive you mad. Fortunately, even when adults try to provide young children with restricted cognitive diets of this kind, children's internal drives are initially so strong that they usually protest and find ways to keep wonder alive on their own, at least for a while.

6 Motivational Muddles

A March 6, 2018, PBS broadcast[1] began as follows: "Many preschool teachers are scared of teaching STEM." One teacher was described as feeling "that children's unrestrained curiosity can sometimes make him feel anxious about teaching certain subjects, like science." In his own words, he said, "I will focus more on reading and not really look at science, because I thought it was complicated." Another teacher said, "I'm not good with the science. I'm just uncomfortable with it." Even teaching the simple mechanics of how objects roll down ramps was regarded with great apprehension by several teachers. One researcher who works in the area, Liesje Spaepen, remarked,

> We hear all the time teachers go into early education to avoid math and science courses themselves. . . . If you're afraid of it, if you have anxiety around a topic, you're very unlikely to want to teach it to your students as well. . . . The problem is, anxieties like that get passed down. If I'm anxious about something, my students see that in me, they think there must be a reason to be anxious about this.

Teachers' attitudes toward science infect the children they are teaching. This influence is well documented by scores of studies.[2] The contagion effects can be painfully nuanced. Children can pick up teachers' sentiments that girls can't do math but boys can.[3] But it is not just teachers who have such feelings. They are widely shared among many adults who at some point in their lives came to regard science as the realm of a special few. Not just in teaching but also in broader cultural practices, adults can undermine children's intrinsic interest in understanding science. Teachers, parents, siblings, and other adults interact with children in ways that reflect their views of science, and, in doing so, they can pass on many negative assumptions about science and technology. These assumptions can also be transmitted by museum designers, children's book authors, toy makers, camp counselors, and church pastors among many others.

How can adults come to avoid topics of such intrinsic interest to young children? Most elementary school teachers were not science majors in college, and many confess to having ambivalent attitudes toward science. They may find science intimidating and doubt their own aptitudes in the area. As one discouraged science educator said to me years ago, young children can "smell their teachers' fear of science" and can't help but become wary of the topic as well. When this message gets echoed at the dinner table, in church, and on the playing field, it can overpower spontaneous interest. Unfortunately, as a result, many children become disinterested and even alienated from science topics.[4] They then complete the cycle of demotivating influences as they grow up and pass their views on to the next generation.

These views are not inevitable consequences of growing up in today's world. We have all encountered adults, both highly educated and not, who share a joyous fascination with science. For many years I lived in an especially rural part of Upstate New York. I used a car mechanic who hadn't completed high school but was deeply interested in science and engineering. He lived on a failed farm thirty miles outside of Ithaca where he fixed cars in a barn. He became good at it and soon was fixing every kind of car, and then boat engines as well. Along the way, he started building his own airplanes, carved a runway out of a pasture, and soon was flying strange contraptions over the hills. Whenever I took one of our decrepit cars to him, I knew I should bone up on recent advances in several fields to be able to carry on a decent conversation. He loved to learn how virtually anything, human made or natural, worked. Somehow, a deep fascination had always remained with him from childhood and seemed to be infecting his children as well.

Our discussion of motivational muddles examines why many children stop wondering in their early school years. We ask how social contexts can erode their questioning tendencies and how adults, while sometimes trying to motivate children to engage more strongly with STEM curricula, can make things worse. What causes some children to lose interest in science and even dislike whole areas of knowledge? Why do some children eagerly continue to drill deeper and deeper into underlying mechanisms while others back off and avoid any such lines of inquiry? Adults can be hugely influential here not only in terms of how they engage with science and technology themselves but also in terms of how they try to motivate children.

Attitudes toward STEM are highly influenced by one's own educational experiences. When preservice elementary school teachers are provided with more content knowledge in areas of science such as biology and physics, they show increased confidence and enjoyment in teaching science.[5] As their content knowledge grows, they gain insight into how to make ideas in physics and biology more pedagogically appropriate for their students. A deeper understanding of content may be just as important in teacher preparation as learning about educational theory and practice and developmental psychology. But teachers are not the primary problem. They merely reflect how many of us view science and technology.

We can develop negative attitudes toward science when misplaced views about children's minds lead to assessments that fuel feelings of incompetence. Much of early science education focuses on learning names, salient properties, and categories. Causal relations may be discussed, but we rarely provide children with anything close to "mechanistic tool kits" for deciphering the world. Instead, causal relations are limited to general ideas such as "animals are adapted to their environments," "animals and plants need energy," and "causes precede effects." Much of instruction dwells on the process of science: measuring, classifying, and sorting. Other curricula stress the importance of "evidence" relentlessly year after year. Many children are deprived of those elements of science that they want to learn about the most, namely what explains a thing's observable properties and why it behaves as it does. When the psychologists Margaret Donaldson, Barbara Tizard, and Martin Hughes described the loss of interest in the school years forty years ago, they repeatedly stressed how most traditional classrooms transform wonder-driven interactive questioning between children and adults into assessment-driven questions.[6] Teachers directed their questions at children in order to check what facts have been transmitted effectively. As a result, questions from children to teachers disappeared. On top of external pressures to assess children, the challenges of deeply engaging in shared questioning, exploring, and discovering with each child can seem overwhelming to many teachers. Ambivalence and anxiety about science makes them feel even more overwhelmed.

The goals of the NGSS (Next Generation Science Standards) have moved in the right directions, but they may not align well with the beliefs and attitudes of many teachers, administrators, and parents. Even when they do align, far too little mechanism may be conveyed. Worse yet, when students' science knowledge is assessed, the tests may undermine motivation rather than nurture it. If an assessment focuses heavily on the recall of facts and isolated relations, it is easy to fail such tests without relentless rehearsal of those items. Even when one does well, the achievement often seems hollow. Memorized facts seem to be of little use and are soon forgotten. In contrast, instruction can focus on enhancing children's already present abilities to learn interlinked patterns of explanatory causal relations. When such explanation-based approaches are used, children gain a sense of mastery as they can use their acquired knowledge to evaluate and build on explanations and to seek out more expert sources. Interconnected causal relations help support each other in recall, enabling more rapid, flexible, and satisfying use in problem-solving and innovation. For example, rather than just list some features of wolves, you might explain how a cluster of features all support hunting prey, such as forward facing eyes and ears, sharp claws and teeth, enhanced olfactory and visual abilities, abilities to run over forty mph for short distances, and strong jaws. A more elaborate explanation might describe how these features causally interact.

When Teaching Goes Well

We all hear about teachers and other adults who are gifted at conveying science content while also inspiring their students. They are a heterogeneous group who work in diverse settings, but there seem to be a few recurring patterns. Consider two exceptional teachers from very different places and schools. One teacher was nationally well-known. The other had an enormous influence on my own education.

The well-known case is Sister M. Gertrude Hennessey, a teacher in a small parochial school in Staughton, Wisconsin. Sister Gertrude, as everyone called her, was the sole science teacher from grades 1–6.[7] She taught science by examining the critical concepts in an area and how they change as children learn. She created with her students a metacognitive narrative that connected with the content. Sister Gertrude saw this metacognitive awareness as helping her students appreciate the evolving nature of scientific knowledge and how it changed through discussion, questioning, and exploration. This appreciation in turn made them value and engage in more discussions, ask more why and how questions, and conduct more explorations to answer those questions.

Sister Gertrude wanted children to clearly state their own ideas and to be able to explain why they endorsed their position. She showed them how to explore the limitations and weaknesses of their ideas and how they might not be internally coherent. She fostered conceptual change through metacognitive insight. This, in turn, changed children's views of knowledge and science. By the sixth grade, her students were talking about the learning of science at the same level as college students. Consider the comments of a sixth grader who had worked with Sister Gertrude for six years:

> In the past I thought for instance the book on the table had only one force, and that force was gravity. I couldn't see that something that wasn't living could push back. . . . However, my ideas have changed. . . . Sister helped me to see the difference between the macroscopic level and the microscopic level. . . . I began to think about the book on the table differently—[last school year] I was thinking on the macroscopic level and not on the microscopic level. This year I wasn't looking at the table from the same perspective as last year. Last year I was looking at living being the important focus and now I am looking at the molecules as being the important focus. When I finally got my thought worked out I could see things from a different perspective. I found out that I had no trouble thinking about two balanced forces instead of just gravity working on the book. . . . now I can see balanced forces everywhere! Balanced forces are needed to produce constant velocity! The book on the table has a velocity of zero, that means it has a steady pace of zero. . . . I have expanded my mind to more complicated ideas! Like molecules in a table can have an effect on a book, that balanced forces and unbalanced forces are a better way of explaining the cause of motion, and that constant velocity and changing velocity are important to look at when describing motion. [grade 6][8]

Researchers who visited the class and Sister Gertrude uniformly insisted that such statements were typical for the children, not just remarks of a few standout metacognitive wizards. Sister Gertrude managed to instill in virtually all her students an explicit awareness of the pleasures of changing one's beliefs. They discovered how change could help them embrace more powerful concepts that enabled them to see new things "everywhere." Children learned that to discover a problem with one's own understanding and correct it was liberating and empowering. They came to understand that finding a problem with their original view was not a failure experience but a great advance.

Sister Gertrude's students also developed remarkable insight into the social dimensions of wondering, as revealed by the following student:

> I think learning in science is about wondering about the world, how it works, and then asking yourself questions that you find challenging. I think a good analogy for learning is a puzzle. You can have all 1000 pieces but if you don't take the time to fit them together you will never see the picture. Most school learning is like collecting the pieces of the puzzle and keeping them in a box. Teachers reward you for collecting enough pieces, the more pieces you collect the better rewards you get!
>
> I think learning in our science class is much different. Back to my analogy of the puzzle. In science class we spend a lot of time trying to fit the pieces of the puzzle together. Sometimes it takes all of us working together to fit one piece of the puzzle into its right place in the picture. The analogy isn't perfect because there is only one way to fit the pieces into the puzzle. But in science there is more than just one way to fit science ideas together. To me, learning is the work you do as you fit ideas together. Using your experiences, talking with others in class, building and testing models that represent your ideas all help you to think about how ideas fit together; and that, I think makes for good learning.[9]

This remarkable teacher's success involved more than a focus on ideas, metacognition, and conceptual change. I had the pleasure of meeting Sister Gertrude when she came to Washington DC to advise the Board on Science Education of the National Research Council. I was serving on that board during a time when we collectively researched and wrote a book on how basic research might be relevant to teaching science in the classroom.[10] Sister Gertrude was completely immersed in the meeting and listened intently to everyone's thoughts and carefully integrated them into her own comments. She had a knack for grasping what others were thinking and where the group as whole might be moving conceptually. She also clearly loved science and the many ways it opened up a wider and deeper awareness of almost any topic. She instantly saw how mundane events, such as that diet sodas float in water and sugary ones sink, could serve as vehicles for triggering conceptual change. In a coffee break, she excitedly told me about how the soda disparity was an excellent way to engage the youngest children in discussion about density. Sister Gertrude was also unfailingly modest about her own knowledge and seemed to regard her own gaps as providing endless opportunities for future pleasures of learning. She saw the act of undergoing conceptual change in efforts to understand a phenomenon as a joyous journey, and her enthusiasm was intensely contagious. This opportunity was open to all her students when given a little guidance on how to travel along that path.

The second example of an inspirational teacher was my science teacher in the early 1960s. Jesse Knight Jr. was the only science teacher in my small school, and he taught the same children from grades 4 to 8. This ability to create and track learning progressions year after year is rare but can greatly enhance a teacher's efforts. "Mr. Knight" (I still feel obliged to refer to him in that way) taught with an ever-present gusto and pleasure. The entire classroom awaited each class as a kind of dramatic performance. He talked most of the time, but we all knew that he might fire a question at any one of us at any moment, and his questions usually asked us to explain a phenomenon or mechanism back to him in our own words. Like Sister Gertrude, he was fully engaged with each student and monitored in real time how well each of us was understanding the underlying concepts. He demonstrated how even a didactic lecture style could keep everyone engaged in a class.[11] This was not staid lecturing from a podium. It was rapid pacing about, with piercing darting glances that made us feel that Mr. Knight could read our every thought.

Mr. Knight's lessons made elaborate use of history. The history may have been embellished or cleaned up at times to make a point clearer, but it worked beautifully as a vehicle for teaching us about the nature of science, about the role of ignorance, and about the importance of communities. I vividly remember how he explained the workings of an internal combustion engine. He started with the idea of chemical energy, how some compounds had energy stored in their chemical bonds. He showed how this energy could be released through reactions that broke those bonds. He described how chemical energy was first harnessed in about AD 1000 in China. People learned how to mix charcoal, sulfur, and nitrates so as to make gunpowder. (This included a humorous side story about guano and animal sources of nitrates.) They then learned how to pack gunpowder into a container and how to rapidly release its energy by heating it and creating an explosion—the first firecracker. He illustrated this in the classroom by creating a bright flash after heating a small amount of gunpowder. He explained that if it were trapped in a container, it could have been a great firecracker, and a dangerous one if the container was made of metal (see figure 6.1).

Mr. Knight then asked us how all this energy could be used in a more controlled way to do work, such as powering a car (thereby introducing the concept of work as well). He eventually guided us to the idea of shaping

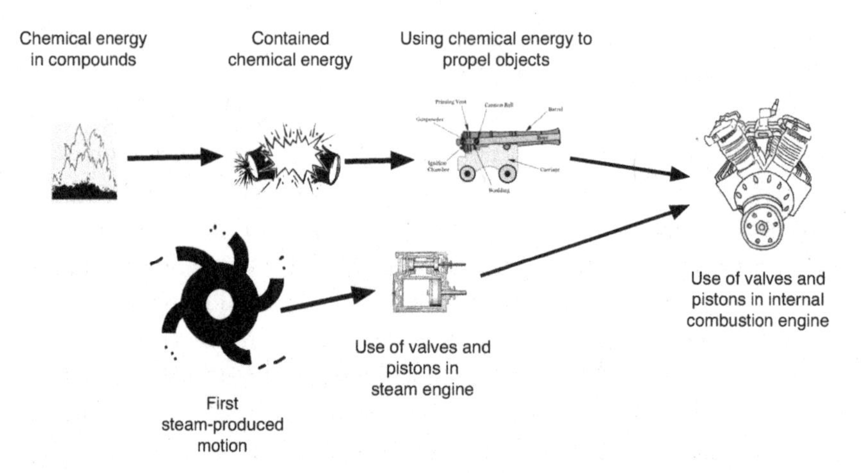

Figure 6.1
Teaching science and engineering through history. Internal combustion engines as taught through the evolution of ideas over thousands of years. The lesson began with the discovery of energy stored in chemical bonds when certain compounds flared brightly when ignited. Containment of these compounds led to early firecrackers and bombs. This led to the use of chemical energy to move objects, such as cannons. This presented a challenge of turning linear motions into continuous rotational motions. To explain how this problem was solved, instruction turned to a parallel history of steam-generated power and the development of pistons, valves, and flywheels. A series of interactions between the two evolving technologies led to the final working internal combustion engine.

the container so that the explosive force was directed. He then drew a cannon on the blackboard. He explained how the Chinese invented cannons and how the idea then spread to the Middle East and Europe. However, he quickly pointed out that you couldn't constantly fire cannonballs in order to drive to the supermarket. There must be a way to control chemical explosions more efficiently. Mr. Knight explained why this was a hard problem that took centuries to figure out. He said it involved a lot of failed attempts, thereby also showing us how failures are to be expected as a natural part of progress.

Mr. Knight then went on a planned digression to discuss how a parallel line of innovations with steam engines resulted in critical insights that were transferred to internal combustion engines in the nineteenth century. He described how people had harnessed steam power even before gunpowder but that a huge advance happened in the early 1700s when people developed the piston. He explained pistons both in terms of their ability to translate linear motion into rotational motion and in terms of a repeated cycle that could provide continuous power (while also showing how heavy flywheels were invented to keep everything turning during low-power parts of the cycle). He described how James Watt and others started using valves that allowed much higher pressures and more efficient movement of steam. He emphasized that the entire process of the steam engine's development took hundreds of years, with lots of clunky models discarded along the way.

Mr. Knight then told us about an individual (perhaps idealized) who, with a piston and valves and using liquid forms of energy such as gasoline and alcohol, made an engine that ran on explosions that occurred inside the piston chamber. He then introduced the piston chamber and valves governing intake and exhaust, the idea of a fuel supply regulator known as a carburetor, and the spark plug. He described all this as accomplished through both trial and error and controlled experimentation. He showed how it involved both cooperation and competition among individuals. He also described lots of failures, sometimes spectacular ones. In the grand finale, he walked over to an actual gasoline engine that was partially assembled under a window on the side of the classroom. He quickly finished assembling it and started it up. It was better than magic.

I have left out many other details. Among others, Mr. Knight explained how piston rings and governors worked, and why cars can backfire and stall on cold days. He somehow did all this in a relatively small number of classes. He taught everything using historical embedding. We learned the history leading up to the first electric motor. He started with discovery of magnetism by the Greeks and ended with us all building small electric motors from kits. He provided historical accounts for biology, covering (1) spontaneous generation to homunculi to genes, (2) bodily humors to bloodletting therapies to tidal theories of blood flow to circulation, and (3) early theories of contagion to microbiology (with digressions about cowpox and milkmaids, and the plague).

I especially remember one class where he used demonstrations to build up more elaborated networks of wonder around atmospheric pressure. After screwing a lid on a metal can in which a small amount of water was boiling by means of a Bunsen burner, he put the can on a cold countertop and continued talking about the history of the first barometer. As we listened, the can started to noisily crumple into itself. As we gaped at the contracting can, Mr. Knight talked about how we were sitting at the bottom of an "ocean" of air that exerted pressure of fifteen pounds per square inch. Our initial wonder now deepened as he challenged us to explain why we didn't all feel squished by this pressure. Each new tidbit of information made our wondering more elaborate and more urgently in need of an explanation. A good teacher knows how to deepen and enrich what we want to know and how to expand our ability to conjecture about different scenarios.

Mr. Knight was certainly the not the first teacher to use historical narratives to teach science and technology. In fact, the Harvard president James Bryant Conant spent much of his later years teaching science through historical approaches that emphasized the process of discovery.[12] Science education scholars have also argued for the value of such an approach at levels ranging from college (most often) to elementary school.[13] Historical approaches have been praised for how they enhance insight into conceptual change, epistemology, the nature of science, and motivation and intrinsic interest.[14] They also interweave examples and "experiences" with principles and models in each new cycle of discovery, much like a current practice of exposing students to the phenomenon first before trying to explain it in terms of principles.[15]

But historical narratives are surprisingly rare in the elementary school classroom.[16] Some argue that all that history takes up too much time in an already overloaded curriculum,[17] but it didn't seem that way with Mr. Knight. History provided a way to integrate content and to make it personally accessible and more memorable. I think he covered content at the same pace as ahistorical approaches by not having to repeat content as many times and perhaps by spending less time explicitly talking about the nature of science or method. Those ideas were already embedded in the historical narratives.

The historical approach may also be avoided in elementary school because of beliefs that one should teach through discovery, exploration, and hands-on activities and not in the form of lectures.[18] However, as the psychologist David Klahr points out, a major gap exists between educational policy and the findings of cognitive science research. While most policies embrace discovery methods of teaching, most cognitive science research in the experimental tradition still finds major benefits to situations in which the teacher spent much of the time talking to students.[19]

Of course, lectures can be excruciatingly boring if they are not appropriately calibrated for the intended students. But when narratives are done well, and this often includes many back-and-forth interchanges between teacher and students, they can be excellent methods of instruction. A second concern might be that teaching through history focuses too much on "dead White males," but this is merely a problem of bad history. History can be a wonderful vehicle for illustrating different cultural perspectives and how people can change over time.

In the end, the pressure to teach the mandated curriculum so that students will perform better on standardized tests may be the primary reason why teachers are nervous about including any extra historical material. Ironically, historical perspectives might even help performance on those admittedly flawed tests. In addition, teachers might need to idealize the history so as not to wade too deep into the complexities of incorrect directions, especially with young children. After all, many early scientific theories, such as earth-centered orbits, were devilishly complicated and only understood by a few adults at that time. Histories for young children may often have to depict more idealized paths that gloss over the complicated detours. They should still make sure children are aware of the detours so they can understand the virtues of intellectual humility, but they don't need to cover detours in such detail.

Why was this approach so riveting and effective for so many of us? In today's terms, it was remarkably inclusive and cross-cultural. It showcased the importance of failures, different and evolving communities of knowledge, metacognition (recognizing bad ideas), conceptual change, trial and error versus controlled experimentation, and effort. Mr. Knight exemplified continuous lifelong learning as he frequently started class waving a new article or newspaper clipping that had changed how he understood something. He often said he didn't know the answer to questions and told us, in those pre-internet search days, how to find out. It was through Mr. Knight's guidance that I made my first trip to the New York Public Library to learn more about the new technology of lasers. Mr. Knight also knew how to blend humor and playfulness with instruction. When students were misbehaving, he would menacingly wave a meter stick and yardstick and ask which one he should use to "whack" them (he never did). He'd quickly explain why they should choose the yardstick because it was shorter and then gave a brief reminder of the many merits of the metric system.

Sister Gertrude and Mr. Knight differed in many ways, but they shared important aspects of approaching and thinking about science. They both exemplified the virtues of intellectual humility and the willingness, even eagerness at times, to acknowledge their knowledge limits. Because of this humility, they saw engaging with science as a lifelong process of continuous learning driven by the pleasure of learning new things, often changing one's mind as a result, and then seeing the world in even greater splendor and beauty. This led organically to a deep love of science content and doing science. They were also deeply interested in the process of conceptual change itself and closely tracked their own knowledge growth and, more importantly, the growth of their students' concepts and beliefs. They both dwelled on why and how questions and often discussed detailed mechanisms. They didn't cognitively condescend to their students and saw themselves as sharing the process of discovery with their students. They saw more commonalities across all students and teachers than differences.

They were unusually driven learners on their own. Both had advanced degrees in other disciplines and were deeply interested in a wide variety of topics. For example, I recently learned from Laura Kang, the current head of my old school who kindly went through the records, that Mr. Knight had secondary degrees in music, mathematics, and education and had studied at Harvard when Conant was there. I further learned that he had become fluent in Portuguese and wrote a technical manual for the Brazilian Air Force. He was also the president of a national organization of science teachers for three years and finally, and perhaps most remarkably, was on the board of directors of one of our country's most prestigious research laboratories. Sister Gertrude was similarly exceptional, including getting a doctorate in philosophy so that she could better understand the epistemology behind conceptual change. These were clearly remarkable individuals who would be rare in any population including schoolteachers, but their passions for learning can be shared with everyone.

Sister Gertrude and Mr. Knight were so exceptional that they may intimidate future teachers rather than inspire them. That shouldn't be the result. Their approaches are repeated daily by thousands of teachers who don't have doctorates or serve on boards of directors. Our three sons in elementary schools in Upstate New York all encountered teachers who helped them think more deeply and inspired new lines of questioning. Sometimes, this took the form of embedding the science in a historical narrative that revealed simple ways to uncover phenomena, such as atmospheric pressure or magnetic fields. On other occasions, it might involve posing a simple question, such as how bats fly in complete darkness, and then gradually unpacking a mechanism through guided how and why questions, accompanied by a metacognitive commentary (e.g., "those are two interesting ideas but they can't both be true at the same time . . ."). In many cases, these teachers had great teachers of their own when they were young and specifically mentioned use of their methods as models.

While we shouldn't underestimate the magnitude of the problems associated with teachers' own negative feelings toward science, those problems are surmountable. Even the most fearful and anxious teachers can come to realize that they share with their young charges a deep-seated curiosity full of unbounded sets of how and why questions. When that curiosity is aroused, it awakens the surprisingly powerful pleasures of insight and discovery, pleasures that can never be taken away and that can be constantly refreshed. Well-worked-out programs for inspiring teachers in this way are rare, but they could be widely implemented. A thoughtful conversation can morph a growing curiosity into a drive to better understand some phenomenon. Wonder is within all of us, awaiting encouragement. Providing teachers at all levels with engaging experiences in learning science content can increase their confidence, their enthusiasm for science, their day-to-day teaching, and their ability evoke fascination in their students.

When Teaching Goes Well at the National Level

From 1960 to the present day, Norway's per capita income and wealth figures have consistently exceeded those of Finland. Yet, in recent decades, Finland has risen dramatically in cross-national assessments of math, science, and language, while Norway has hardly changed despite much higher levels of spending per student.[20] Finland's educational progress was more apparent when the PISA test, the Programme for International Student Assessment, became widely used in many countries worldwide. PISA, which is given every three years to fifteen-year-olds, assesses not only stored knowledge but also the ability to extrapolate knowledge to unfamiliar situations and types of problems. Finland's PISA scores put it among the top countries in the world, far above countries such as Germany; the United States; and its Nordic neighbors Norway, Denmark, and Sweden, most of which far outspent Finland in funds per student year after year. Over the course of several PISA cycles, Finland has largely maintained its standing among the highest ranking countries, most notably in science.[21] Moreover, Finnish scores have some of least disparities among socioeconomic and ethnic groups. Students from all backgrounds are thriving.

The improvements in Finland's schools became a hot topic for the general public when the popular documentary *Waiting for Superman* contrasted education in Finland to the dismal state of education in the United States. Hundreds of education delegations from other countries have made pilgrimages to Finland over the past decade to understand what happened that transformed the country. What did they find? Finnish officials had deliberately decided to change the educational culture in fundamental ways, much of which centered around inculcating new values for learning, for teachers, and for knowledge itself.

One of the most dramatic changes was a transformation of the teacher education system from virtually open-admission teaching colleges into ultra-selective programs in universities. These university programs currently accept only 10 percent of applicants. The journalist Amanda Ripley described getting accepted in one of those schools as being as difficult as getting into MIT.[22] Successful applicants have outstanding high school records in contrast to admitted student teachers elsewhere. As the schools metamorphosed into centers of excellence, teaching became a highly valued and respected profession. Teachers were compared to doctors and lawyers in terms of training, ability, and status. Pay went up, but it was not the primary driver as highly paid teachers in other countries from less selective colleges did not have the same successes in their schools.

As the entire country showed an increased appreciation of the value of learning, students didn't need to be cajoled and constantly prodded into studying as occurs in some other elite scoring countries. Instead, they internalized the growth of knowledge as a highly valued outcome.[23] Ripley describes how even the most conspicuous "stoners" in Finland's high schools immersed themselves after school hours in challenging books and materials related to their academic interests. When asked why, high schoolers seemed confused; to them it was like asking why you should eat meals. Why wouldn't they want to expand their knowledge and understanding in a topic they cared about and where new insights could be so rewarding? Self-driven learning and the idea of exploring intriguing challenges were universally valued. Standardized tests were eliminated as the country trusted the teachers to best monitor and address each student's progress.[24] Their trust was well-placed. Teachers tracked each student's progress through extended one-on-one interactions. In many ways we see in their interactions with students echoes of what made Sister Gertrude and Mr. Knight so effective.

Was the elevation of teachers as gifted professionals the critical change that led to a cascade of other effects? This interpretation underlies many reports from within Finland and by outside observers. The teachers are praised as the best of the best, winnowed out by selective admissions and rigorous training in elite teacher training institutions. They are also described as knowing each child extremely well, often tracking them cooperatively with other teachers and with support from excellent teacher's aides. When a child starts to lag behind others, they quickly ramp up resources to help the student catch up, descending on students as pedagogical SWAT teams. Teachers seem to "inhabit" their students' minds, often roaming the classroom and talking to each child about his or her current activity.[25]

Finland's schools differ in other ways as well. Standardized tests designed to make schools and teachers accountable simply don't exist while they have proliferated in Norway. Few curricular guidelines exist beyond a shared mission of building intellectually curious, inquiry-driven minds. The bureaucracy is minimal, with all principals serving as teachers as well. But for many observers, these other differences may emerge as a consequence of developing a new corps of outstanding teachers. Teachers were compared to doctors, making diagnoses of learning challenges for each child and treating them effectively so that the child moved forward. Teachers became trusted professionals who knew best about how to devise individual programs of study with children and how to monitor their progress. Control of curriculum and assessment flowed naturally from the central authorities to local schools and individual teachers. This new corps of teachers also explicitly embraced many of Dewey's ideas of student-centered learning.[26]

Most observers praise the Finnish approach in comparison to that taken in countries like Korea. Those other countries achieve high PISA scores but involve many more hours of schoolwork and homework, intensive testing, and additional coaching and tutoring. They are widely seen as creating aversive experiences for students and as being less likely to build an intrinsic love of learning.[27]

Could an educational transformation like Finland's be achieved in a large country like the United States? It would be challenging if it required shutting down traditional teacher's education programs and creating highly selective new programs. Could the United States meet the need for hundreds of thousands of new teachers each year, especially given that applications to teaching colleges have been declining?[28] Could lawmakers and the general public be convinced to fund such changes? If the teachers are the critical factor, do they have to be intrinsically gifted and then superbly trained? Or could less initially gifted students become great teachers by guiding them to Finnish-style educational values and practices and by convincing the broader public that teachers deserve the same respect as practitioners of other elite professions?

We don't know the answers to any of these questions. One guru of Finland's education success story, educator and writer Pasi Sahlberg, has argued that the most important change is the country's attitude toward education and learning. The public must recognize both the intrinsic value of learning and regarding those who teach as members of a noble distinguished profession.[29] When that happens, applications to elite teaching programs will presumably soar as applicants see future selves as respected, well-compensated members of their communities. Compensation is not the whole story, but it indicates respect for the profession and promise of a reasonable lifelong career. When teachers went on a statewide 2018 strike in Oklahoma, they listed respect for their work as a critical problem for which their lower salaries were just one painful indication. Some argue that you can't elevate teachers to such distinguished positions without first developing elite programs producing phenomenal teachers.[30] In reality, the two would probably have to go hand in hand in an upward spiraling cycle. After all, it took more than ten years for Finland's changes to show major effects.

However, a change of attitude will not be easy. As Finland fine-tuned its system, in the 1990s, many other developed countries launched accountability movements that compared outcomes across schools and teachers. They ramped up testing and announced "races to the top" and other competitive programs. Most teachers saw these changes as downgrading their profession and as demoralizing. Moreover, not only were external incentives alienating teachers but comparable efforts with students were also employed, often with even more counterproductive effects.

The Play/Wonder Launchpad

We cannot count on massive education changes to occur at the national level, but we can focus on play and wonder in preschool and the first years of formal schooling as forming an early launchpad for later more elaborated wonder. Much of Finland's success involves embracing children's intrinsic curiosity and playfulness. Finnish educators stress the vital centrality of play, which is often manifested as wonder-driven exploration and discovery.[31] If teachers in other countries, regardless of prior training and selection processes, downplayed their own anxieties about science by embracing shared playful inquiry in a cooperative setting with their students and their colleagues, more deep conceptual learning would occur. We all have large gaps of knowledge and should welcome the process of posing why and how questions to address those gaps. We then experience the pleasure of gaining new insights. Teachers would see more clearly the vital importance of closely tracking each student's progress. They would be more able to do this if their judgments of student progress were valued and the multibillion-dollar testing industry was eliminated. This change would reduce many motivational pitfalls, but, for it to be effective, entire communities of parents, teachers, and schools must cooperate in the endeavor.

Wonder is often intertwined with play, especially in early development. Play is extensively discussed elsewhere,[32] and a detailed account is not feasible here. However, a few key points need mention. First, play dominates descriptions of the ideal motivational context for learning. In Finland's educational system, play is deemed essential, especially in early school years. Indeed, Finnish children do not have to start formal school until age seven because play experiences in most homes or neighborhoods are seen as offering unique learning opportunities that might be stifled by even the best schools. Then, in the first years of school, instruction is considered subordinate to play. Teachers do strive for "deeper play," where teachers constantly identify and provide opportunities for play that will challenge children by posing fascinating problems and expanding their understanding.[33] They recognize that aimless unguided play can fail to provide rich learning experiences. Teachers continue do whatever possible to preserve the vital motivational ingredients of play right through the final years of high school.

Second, because play appears so early and spontaneously in the preschool years, it does not have to be taught or encultured. It merely has to be appreciated and understood. An adult who appreciates the immense cognitive value of play and how it can be linked to creativity, problem-solving, reflection, and self-determination should support what is naturally present and help it grow. Sadly, all too many assessment-driven education systems ignore the motivational and cognitive benefits of play and see it instead as a frivolous activity to be eliminated. Some schools even outlaw play in kindergarten as they attempt to push teach-to-the-test methods down to lower and lower grades.

Third, play's early emergence also explains why play and wonder are cross-cultural universals and were present everywhere long before formal schooling existed anywhere. Many anthropologists have observed that children in traditional cultures often continue to show extensive play through adolescence. Archeological records show the presence of many prehistoric play versions of adult artifacts, ranging from miniature bows and arrows, to toy boats, to clothed dolls. One theory proposed that, in prehistoric cultures, play with such toys during adolescence was a "sweet spot" period in development where a surge of adolescent innovations in toys led to improved mature versions.[34] Such cases are not idle play but are infused with wonder. Preservation of such activities in older children and even adults may occur more often in many traditional societies even today. Perhaps those adults who are so skilled in lifelong playful wonder, if introduced to modern science and technology in appropriate ways, could be especially fertile sources of new ideas and designs.

Fourth, play is amplified by social interactions. Play with others creates new opportunities for experimentation, exploration, and reflection. Play is a superb setting in which back-and-forth wondering can thrive. In Finland's schools, deep play is described as occurring in contexts of shared questions and speculations that exemplify wonder, such as "H'mm, I wonder why that does . . ." or "Can we figure out how to . . . ?" While play and wonder are clearly distinct and each can occur without the other, they reinforce and amplify each other in marvelous ways.

Mathematical Tangents

Interviews with elementary school teachers often uncover feelings of anxiety about mathematics that factor into anxiety about science as well.[35] That anxiety can be amplified when those with weaker math backgrounds take courses in physics and then extend this apprehension to all the sciences. A full discussion of learning mathematics is far beyond our scope and is an area that is rapidly evolving in cognitive science, including studies in our lab suggesting new ways of understanding the origins of mathematical cognition in infants and young children.[36] In addition, most math anxiety studies have involved college students. However, a growing group of studies find early signs of math anxiety. By first grade, some students have math anxiety and the attendant negative effects on math performance.[37] For young schoolchildren, some of the best interventions involve providing more math. When children are exposed to an increase in math materials through skillful one-on-one interactions, they start to realize that they have no good reasons to be anxious given their unmistakable progress.[38] Helping children reappraise their own efforts and abilities can improve learning results in all of science but also more specifically in math. Here, three key points are especially relevant to links between science and math.

First, when math is taught and assessed using diverse techniques, it can be far more inclusive in terms of achieving major gains in understanding across diverse populations. Teachers can use techniques for building math concepts that don't disadvantage those who are not as adept at numerical calculations. Assessments do not have to focus just on calculation speed.[39] For example, children might practice solving magic squares of increasing complexity with no time pressure and through trying out relatively simple addition patterns.

Second, large increases in math understanding involve discovering its rich internal structure. In elementary school, I was amazed when I first learned that no map ever needed more than four colors to ensure that no adjacent countries or states were of the same color. I was even more amazed to learn that, at that time, no one had yet figured out how to prove it must be true. Like many other young children who learned these two facts, I drew hundreds of maps thinking I might find the exception, always trying to sneak in slivers of imaginary countries to break the rule but never succeeding. (The four-color theorem was proved in 1976 by Kenneth Appel and Wolfgang Haken.) A little later, I learned that, while I could easily bisect an angle with a compass and a ruler, it had been proven over a hundred years earlier that it was impossible to trisect. Because the proof itself was challenging to understand, I stubbornly tried to find new trisection techniques for over a year before surrendering to authority.

A few years later, I started to learn about the mysterious world of pi. I learned how certain series of additions and subtractions of fractions can converge on pi. This is an extraordinary thing to encounter. You can know the precise ratio of any circle's circumference to its diameter by calculating certain infinite series: ($\pi = 4(1 - 1/3 + 1/5 - 1/7 + \ldots)$ or ($pi = 3 + 4/(2 \times 3 \times 4) - 4/(4 \times 5 \times 6) + 4/(6 \times 7 \times 8) - 4/(8 \times 9 \times 10) + \ldots$). Even more surprisingly, the digits that make up pi have been calculated to many millions of decimal places and appear to come in a random order, but after thousands of years, no one has been able to prove that the sequence actually is random. It still makes my head spin.

Each year of additional math instruction reveals new beautiful hidden structures. New insights emerge when you understand calculus, envision families of differential equations, and appreciate the dimensional nature of linear algebra. Math ceases to be seen as a set of calculations and more as a system of structures to be discovered, much as science discovers causal and mechanistic patterns. Mathematical relations are not causal, but in many other senses, math exploration can resemble areas of science.

Third, even if you aren't fortunate enough to have had a strong mathematics background, you can still learn about traditionally quantitative areas of the sciences such as physics. Profound insights into physical systems can be conveyed to students with diverse math backgrounds. Some of the greatest discoveries in science, such as much of Einstein's theory of special relativity, can be understand with knowledge of simple algebra. Others may require no math at all. Some science teachers have adopted a "conceptual physics" approach in which they convey important physical concepts with a minimum of mathematics.[40] Concepts such as Newton's laws of motion can be initially conveyed and demonstrated without numbers. "Objects at rest will remain at rest, and objects in motion will remain in motion at the same velocity, unless the objects are acted on by an external force" and "When one object exerts a force on another object, the second object exerts an equal and opposite force on the first." Concepts of work, conservation of energy, and momentum can also be conveyed in nonquantitative terms as can some aspects of the idea of entropy. Grasping mechanisms in many areas can be largely orthogonal to sets of equations.

All students should learn that mathematical descriptions can be enormously helpful in discovery and should be provided with rewarding ways of growing their mathematical skills, but no one should feel that they cannot gain valuable insights into nature and technology until they fully master mathematics. That said, it may be unfair to attack rote learning of times tables and other math facts, or *math fluency*, as creating math anxiety and destroying motivation and to advocate a nearly exclusive focus on meaningful math use, or *number sense*.[41] Mastery of facts can happen in a fun and painless manner with all children. It doesn't have to be what critics call "drill and kill." When sufficient fluency is achieved, the subsequent liberation of working memory space allows children to then see patterns and more complex math principles, which can both be fascinating in their own right and provide motivation for learning more math.[42]

Math anxiety affects math learning worldwide. Across all the PISA countries ranging from those that score the best in math to those that score the worst, higher math anxiety in almost every country strongly predicts poorer mastery of math.[43] This result shows just how much math learning is influenced by anxiety in all groups, usually through anxiety in parents and teachers that is communicated either directly by negative and fearful comments about math or indirectly through behaviors indicating anxiety and distaste for the topic. Even in the top-ranked math countries that emphasize cooperation and few grades and tests, such as Finland, greater math anxiety predicts worse math performance.[44] Anxiety's influence on math requires its own kinds of corrective interventions regardless of any level of actual ability.

Relevance and Attitude

Conceptual physics brings up an associated motivational issue concerning intrinsic interest in science and technology topics. A child might claim, often based on a familiar adult's attitude, that he doesn't care about physics, or biology, or even science in general. He might declare that it is boring or not worth the effort of trying to understand. He might also insist that he has nothing in common with others who like those topics. I remain skeptical about all such claims given how virtually all young children tend to be omnivorous learners. Often, such objections can be overcome by making the discipline relevant to something important in their lives. For example, if a student loves the visual arts, one might show how the physics of light can enhance their ability to understand the creation of perceived colors by an artist. Many great artists were keenly interested in relevant principles from physics. Painters such as Caravaggio, Leonardo, Vermeer, and Monet carefully studied the nature of light to better understand how to create certain effects.

Any discussion of negative attitudes toward science is likely to be bring up questions of gender. Summaries of the large literature have considered everything from prenatal exposure to androgens and spatial ability to stereotypes.[45] Those reviews inevitably conclude that sociocultural factors, such as beliefs about differences in ability or about gender-appropriate life roles, are by far the strongest predictors of attitudes toward science, majors in college, and career choices. Because cultural and societal effects extend far beyond the family level, especially as children enter the middle school years, making changes that will provide girls and boys with a wider range of future choices requires far more than ideal intentions and behaviors in parents, but such intentions are a great place to start as parental attitudes early on can have long-term effects.

Framing of how the self is related to science can moderate sociocultural influences. One surprisingly simple manipulation is to talk about doing science rather than being a scientist. When girls between the ages of six and eleven were asked about their interest and potential to "be scientists," they showed a characteristic drop in interest and feelings of efficacy as they grew older. In contrast, they showed much less of a drop when asked about their interest and potential to "do science."[46] Apparently, doing science was perceived as more inclusive and inviting than being a scientist. Being a scientist may have implied some degree of essential science ability as opposed to an activity in which all could engage. This encouraging finding suggests how relatively simple changes in descriptions of science learning can influence motivations and views of future careers. Using language in this way, however, may also pose risks if not done carefully. If "doing science" is implemented as a curriculum with more emphasis on the procedures of science, such as measuring, formatting of data, and strategies for controlling variables, and less emphasis on models and mechanisms, it runs a serious risk of reducing interest with all students. Most people would prefer to learn how things actually work than how to most effectively collect and confirm evidence for those models and mechanisms.

Ruinous Rewards

Other motivational misunderstandings arise from widespread beliefs about rewards. In schools and many other settings, prizes and rewards are considered obvious ways to enhance interest and motivation. Yet, decades of research have shown that positive well-intended rewards can sometimes undermine motivation and interest. This negative effect may seem paradoxical. We all have heard about experimental participants ranging from pigeons to pediatricians whose behaviors were shaped by appropriate schedules of positive or negative reinforcements. How could encouragement of a behavior destroy intrinsic interest in it?

Almost fifty years ago, Mark Lepper and colleagues at Stanford published a series of papers showing how preschoolers could be led to avoid what they originally loved.[47] The experimenters invited children into a special room in their preschool where they had not been before. The children were assigned to one of three conditions: The first group were told that would receive a "good player" award if they used a set of novel markers to draw nice pictures (all children in this group were rewarded). The second group were simply invited to play with the markers and were unaware of the reward while drawing the pictures, but rewards were given to all of them as a surprise at the end of the session. The third group was just like the second except they never received or heard about the reward.

The first group of children showed a large drop in motivation. Those hapless children who initially quite liked using the markers and knew that they were working for a reward soon came to avoid the markers when they were later made available in their classroom. In contrast, the surprise reward children kept using the markers at a much higher rate, the same as the third group. It wasn't the presence of the reward that reduced the pleasure; it was *knowing* that one had engaged in the activity with the very salient reward in mind at all times. This is often described as children reattributing the motivation for their activity to extrinsic incentives instead of to intrinsic interest, or as Lepper put it, "turning play into work."

These early studies are well-known in the psychological research community, and many follow-up studies have uncovered nuances concerning when and how intrinsic motivation can be corrupted and when it is not.[48] For example, if the task is initially not enjoyed, rewards may increase intrinsic interest as children discover it wasn't quite as bad as imagined. One might therefore link rewards to less pleasant parts of a science class not central to learning science, such as the need to keep the school's textbooks clean and unblemished. But when the task has preexisting high levels of intrinsic interest, rewards can be destructive. Unfortunately, in schools and other settings designed to teach children about science, the contexts can be ideal for undermining interest. Highly salient rewards are discussed extensively at the outset of the school year and are often omnipresent during learning. When the learning period is over, there may be elaborate ceremonies celebrating those children who performed best. By adopting a mistaken view of how rewards link to motivation, well-meaning adults may be undermining children's intrinsic interest.

Perhaps even worse, perceiving oneself as working toward a reward can hinder creativity. Both children and adults may start to view a task as piecework drudgery instead of a pleasurable activity. In tasks with a salient reward, they may churn out many more products, such as piles of drawings, but of uniformly poor quality and showing less creativity as rated by others[49] (see figure 6.2). Big prizes can work against students coming up with the most creative projects and can discourage participants from joining similar contests in the future. If prizes are a tradition that can't be stopped, those giving prizes must emphasize that creativity, not amount of product, is being rewarded. They also should try to instill agency and a sense of self-determination by giving contestants choices among several rewards and among tasks.

The dominant pattern from the early elementary school years into middle school is a continuous decline in intrinsic motivation to learn new materials in school, with math and science often showing the sharpest declines.[50] In many schools, an ever-present parade of extrinsic incentives coincides with environments that discourage individual autonomy and agency by presenting gauntlet after gauntlet of high-stakes tests. Many studies have explored ways to reduce the toxic effects of rewards and even, in some cases, to make them motivation enhancing. But for adults who have not been immersed in those research literatures, the most commonsense uses of rewards can often be counterproductive. At the very least, adults should learn the potential pitfalls of using rewards to motivate children. As the education author Alife Kohn wrote, "Do rewards motivate people? Absolutely. They motivate people to get rewards."[51]

An alternative remedy to the undermining toxicity of extrinsic rewards is to embrace young children's interest in science. A sustained history of intrinsic spontaneous interest in science and technology and in discovering inner workings may counter newly introduced external rewards. If a child repeatedly enjoys the activity in its own right, rewards are less likely

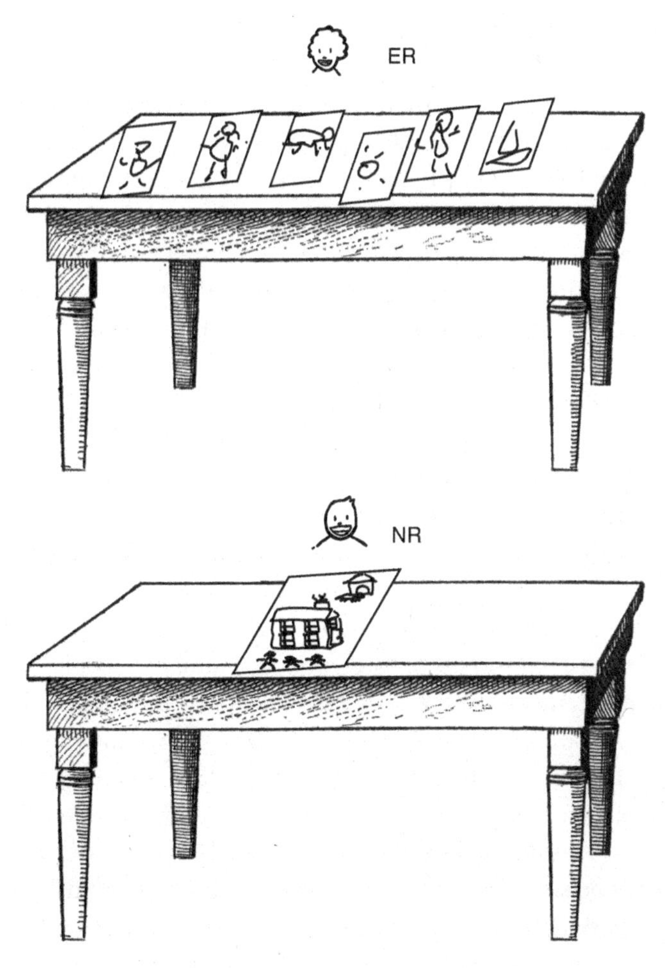

Figure 6.2
The corrosive effect of extrinsic rewards on creativity. The expected reward (ER) child, focusing on mere quantity, churned out a large number of shoddy drawings, while the no reward child (NR), drew a much more aesthetically interesting single drawing.

to have an influence. Moreover, if through such experiences, one has a well-developed narrative for why one intrinsically enjoys an activity, a reward won't trigger a reinterpretation of the reasons one is engaging in the activity. In some cases, the reward can even be construed as pleasant bonus largely unrelated to one's primary motivation. Thus, prior rewarding discoveries may well be the best way to immunize oneself against other misguided material rewards.

Multiple Mindsets and Developmental Destinies

A highly influential view of motivation and learning favors incremental over fixed mindsets. As articulated by Carol Dweck, some people see their cognitive abilities as "fixed" and unchangeable.[52] Either one is good at math or one isn't, and no amount of experience will change one's basic ability in a major way. Learners with such a mindset find it hard to persist in the face of initial failure experiences. In contrast, learners with incremental mindsets see their abilities as malleable through effort and practice; they have the desirable "growth" mindsets. They believe that something like math ability is not an innate essence, but more like a muscle that can be strengthened with dedicated practice. Many studies have tried to induce incremental mindsets in children, especially those who have the most fixed views of their mental abilities.

Some studies have achieved short-term increases in incremental mindsets by praising effort rather than ability.[53] Other studies find long-term relations between parental behavior and mindsets. For example, in one study, the extent to which parents praised effort in young preschoolers predicted more adaptive and resilient motivational orientations in those children five years later.[54] Praise of effort could be as simple as saying "You work so hard." Praise for fixed traits was often described as "You're so smart." Adaptive and resilient motivational "frameworks" characterized those children believing in trait change and having preferences for learning over performance goals. They persisted even when they performed poorly at first. The long-term effects were impressive, but the parents were not randomly assigned to groups and instructed to praise in different ways. Instead, the researchers tracked the extent to which parents normally engaged in one form of praise or another, a practice that likely continued for much of the five-year period.

Achieving success with interventions on a large scale remains a challenge. One notable study addressed mathematics instruction in lower achieving secondary school students across the United States. A single hour-long growth mindset session was able to improve later grades.[55] But replications of that success remain elusive. One analysis of twenty-nine separate mindset interventions found that only 12 percent had positive effects on academic achievement, and those tended to be small.[56] Even the much more modest finding of a positive correlation between mindset and academic success happened only 37 percent of the time. Thus, setting interventions aside, a child's natural mindset is not strongly related to academic achievement.

Incremental views, however, may be encouraged more easily and naturally by taking advantage of a robust developmental phenomenon discovered in a series of studies by Kristi Lockhart.[57] Young children are markedly more optimistic about their future abilities than are older children and adults. Throughout the world, young schoolchildren see almost no limits to their future abilities, whether those abilities reflect physical attributes such as height and strength, social attributes such as leadership, or intellectual abilities such as math skills.[58] After age ten, a decline in optimism about future abilities occurs. This drop may help explain why depression surges in adolescence as teenagers start to conclude that their negative traits might not ever change for the better.[59] Other studies in our lab also show that children between the ages of five and ten are strongly optimistic about their ability to acquire knowledge.[60]

With respect to attitudes toward science, the early emerging optimism bias may instill growth mindsets toward science. The more a child experiences informal learning about science as a preschooler, the more that child will see science knowledge as an incremental ability. When children experience scientific understanding as steadily improving over several years through hundreds of small advances, they are more likely to maintain that attitude in the face of countervailing pressures when formal schooling begins. A child who learns the value of arguing to learn, of using mechanistic information to build abstract models, and of asking why and how questions not only will naturally experience rapid growth of knowledge but will envision a future frontier of exciting explorations.

At a young age, we may not need interventions that explicitly try to inculcate incremental mindsets. Instead, we can grow up in environments that nourish children's fascination with how the world works. The best interventions may be those that preserve this early delight in discovery and insight. An incremental or growth mindset may automatically follow from such positive learning trajectories and from having continuously active PHED cycles (puzzling, hypothesizing, exploring, and discovering). This might be especially true if children become metacognitively aware of their growing knowledge. We can help children better appreciate all that they have learned and not discourage them with factoid assessments that overlook these more impressive achievements. Sometimes, a child may be so intrinsically inquisitive that the child maintains a high state of wonder throughout life despite growing up in environments that discourage this way of thinking. The child's drive to know provides a vaccine against epidemics of negative influences. Much more often, even these children need a wonder "booster shot" from at least one inspirational teacher or mentor. Some combination of these factors may explain why my flying car mechanic retained his fascination with science and technology.

Beyond Mere Curiosity

Children, and many other young animals, have been frequently described as especially curious, but the meaning of curious varies widely across authors, who often decry the variability in meaning and offer their own sense.[61] Some meanings describe curiosity as a basic drive like hunger. Others describe it as a specific need to fill in gaps in knowledge.[62] Nonetheless, most researchers invoke information-seeking behavior. Curiosity involves finding bits of knowledge to plug a gap rather than uncovering and tracking down a rich explanatory structure. This is why wonder is more than curiosity. Wonder contemplates different kinds of future paths of knowledge exploration and construction. For this reason, when researchers describe declines in curiosity during elementary school, declines in wonder are often implied but rarely made explicit. The psychologist Susan Engel comes close in an essay on the decline of curiosity:[63]

> What might account for the difference between the kind of exploration and question asking noted in toddlerhood and preschool and the stunning lack of such inquiry in school-aged children? What happens during childhood to prune away so much spontaneous investigation and eagerness for new information? The answer lies in part, once again, with the adults. . . . Kind and skilled teachers often unwittingly and subtly push inquiry aside. . . . In an effort to meet current state and federal standards, many public schools are consumed with training children rather than educating them. . . . Young students spend their days identifying letters, reciting written words, answering specific kinds of questions, and enacting routines. Older children spend their days practicing specific academic formulae, rehearsing information, and learning how to follow written instructions.

Engel's description of declining curiosity is echoed by many others who provide accounts implying wonder.[64] The decline also overlaps with reasons offered by researchers to describe drops in intrinsic motivations to learn. As children transition from those exuberant preschool days of eager exploration and joyous discovery, we must find ways to preserve and nourish that way of encountering the world. All children should experience what apparently only a lucky few adults, at any level of society, experience today.

How you argue can also have strong effects on motivation. Children and adults alike thrive when they focus on why and how questions and when they argue to learn instead of to win. Win-lose framings can insidiously steer a child toward ineffective and destructive uses of arguments and quizzes about facts where the winners may be declared more easily and quickly. In thousands of ways, subtle shifts in such contexts can either promote natural and joyous learning or make the whole experience stressful and unappealing.

Arguments to learn are intrinsically social and mutually rewarding. Children enjoy these kinds of interactions, especially when guided by a gifted teacher who asks open-ended questions, prods students to clarify what they are actually studying, focuses on interconnected causal relations, and helps students develop clearer understandings of what they do and do not know.[65] Teachers can do all this with young school-age children, as both Sister Gertrude and Mr. Knight did, but this set of discussion skills takes time to develop and may require extensive mentoring by teachers.[66]

■ ■ ■

Motivational factors should not be underestimated. They can turn play into work, turn productive discussions into ad hominem attacks, and lead to views of the self as intrinsically and forever less competent. Many ways to address these concerns are actively being pursued. Naturally emerging early patterns of curiosity, discovery, and spreading insight, when sustained and nurtured, can be powerful forces in their own right against many of the motivational traps that can ensnare children during the school years. This happens when fully involved communities of adults, including parents, teachers, and many others, agree on preserving a certain frame of mind—a frame that continuously expands children's abilities to see and understand causal patterns all around them. Equally important are those exceptional adults who naturally and playfully model wonder throughout their entire lives. Play is difficult to experimentally investigate, but we all know how it can be an extraordinary motivator in its own right. We have also explicitly discussed how easy it is to turn play into work. Play is often at the heart of intrinsic motivation. Adult models who are both playful and also passionate about expanding their understanding may be the most compelling models of all. Dewey knew this more than a century ago when he said that no one who aspires to the profession of teaching will ever be successful in the long run if that person doesn't always retain a passion for learning and deep intellectual curiosity for his or her entire career.[67] His description of deep intellectual curiosity is a version of wonder. Wonder is not casual speculation. It is a driving quest to know and understand.

Unfortunately, these enduring positive attitudes and motivations may be rare, and, instead, negative influences create teenagers who are less able to recognize and override distorting biases, more susceptible to community pressures, and generally more disillusioned and negative about science. Moreover, most children start down this discouraging descent in the early years of schooling. These negative influences can continue to distort cognition throughout our adult lives.

III Disengagement and Its Discontents

7 Seductive Detours

It is rarely a good idea to ask people to estimate their competence in an area outside their own sphere of expertise. I recently talked with a homeowner who had learned that his entire septic system needed to be replaced. He was sure the costs were modest; all one needed was a pipe leaving the house that then branched out into many other pipes into a "septic field." He had roughly calculated the cost of PVC pipe at a local discount store, added in a few hours of shovel work, and came up with the "generous" estimate of "$2,500." The lowest estimate from three reputable companies came in at $43,000. He was convinced that a massive "septic system conspiracy" was at work.

How could such a simple configuration cost so much? The homeowner had grossly underestimated the complexity of septic systems and how they actually work. While septic installations are not supersophisticated, they do involve a fair number of causal considerations. The state of Connecticut "design manual" for residential septic systems runs more than 163 singled-spaced pages in which theory and practice are intertwined. No regulations are cited, just page after page of text and complex diagrams about design of the overall system and its components. You must know the typical constituents of residential sewage; the "bio-chemical oxygen demand" of a typical household; the roles of nitrogen, coliform bacteria, and phosphate; the effects of hazardous chemical "additives" and other atypical things that get into systems; and usage patterns of typical households, among many other considerations. You need to thoroughly understand the site in terms of soil composition and geological features. You must know how a "perc" (percolation) test is done and why it is done and all the factors such as time of year that can distort readings. Each of these steps involves many substeps. After reading this guide, the costs of replacement are much more understandable. How could the naive impression be so far off?

When mistaken views of children's minds and motivational missteps undermine wonder during childhood, what are the long-term consequences for adults? In many instances, they become more susceptible to two kinds of distorting influences that continue to hamper understanding long after childhood. The first is the robust continued presence of early naive theories and models of the world. These may have served as useful convenient fictions for initially thinking about phenomena, but when they endure full strength in adults as *lingering legacies*, they distort judgment and causal reasoning. The second involves illusions of understanding in which people overestimate how well they can explain things. When people are unaware of gaps in their knowledge, they don't appreciate the need to develop that knowledge further. These two influences can short-circuit deeper thought about mechanisms. An awareness of these influences can help counter them, but we often don't see them at all. Instead we may blithely coast along thinking we understand the world quite well. These illusions of insight and understanding can breed a cognitive complacency. We then end up not only being stunned by septic system repair estimates but more generally being trapped in seductive detours from the truth that lead us off the path of greater understanding.

We can resist these detours when we continue to engage in thoughtful questioning, conjecture, and exploration. Instead of tolerating the inevitable inconsistencies that come with these legacies and illusions, we can ask what alternative accounts might better explain phenomena in a more coherent manner. This is what happens with a well-nourished and supported sense of wonder.

It might seem we can't possibly free ourselves from these detours. The world is too complicated for our limited minds, and we will have to settle for highly distorted gap-filled impressions while believing that we in fact know a great deal and with high fidelity. But such pessimistic accounts are unwarranted. The extraordinary abilities of young children to track real causal patterns, to leverage deeper understandings in other minds, and to appreciate the special power of mechanisms tell us otherwise. If we continue to expand on those early capacities, largely though engaging in deeper and deeper cycles of wonder, we can keep those detours in their place. That more optimistic view is reinforced by reflecting on a small group of historical figures who seemed to defy all such constraints, those alleged polymaths who seemed to know everything there was to know. A closer look at these people and what they actually did know at different points of history reveals habits of mind and ways of wondering that are open to all of us, not just those rare outliers with extraordinary amounts of knowledge. We can all learn how to manage our compelling, but wrong, "gut" beliefs and our tendencies to succumb to illusions of insight.

Lingering Legacies

Children may use "convenient fictions" to bootstrap their understanding of the world into more mature forms. An essentialist bias could prompt young children to go beyond surface appearances and look for underlying features and inner mechanisms.[1] That sense of something beyond the obvious supports more sustained investigations. The essentialist bias can work without any prior knowledge of the category itself, especially with naturally occurring categories, or "natural kinds."[2] It simply directs learners to focus on an entity's interior and its microstructural properties.

But an essentialist bias hinders an understanding of the evolution of new species through natural selection. It implies that all members of each species must share the same fixed essence. Fixed essences contradict the reality of species as distributions of genotypes and phenotypes with no species-defining genotype shared by all members. Subsets of these distributions are *selected* when organisms in those subsets fare better than others in specific environments.[3] In an ideal developmental trajectory, a child might first benefit from an essentialist bias to "dig deeper" in efforts to make sense of things. The child would then gradually inhibit that bias as he or she learned about evolution. In turn, suppressing the bias would reduce the tendency to falsely essentialize aspects of other categories such as gender and race.[4]

Essentialist vestiges can even influence our views of scientific discoveries versus revisions. Studies in our lab showed that the more a category is essentialized, the more discoveries are embraced and revisions are rejected.[5] We also found that the essentialist bias varies in strength depending on whether you are thinking of a category in natural terms (e.g., biological properties) or in social terms (e.g., cultural practices).[6] For example, you are more likely to essentialize properties attributed to a group described by a biological feature (e.g., The Vawni people have freckles on the bottom of their feet) than by a social one (e.g., The Vawni people worship the sun as their god). This contrast holds in young children, indicating early conceptual flexibility with the essentialist bias, and suggests a lever for modifying essentialism.[7]

Early ways of understanding can persevere in adults even when they are patently false. The psychologist Andrew Shtulman describes how practicing astronomers take longer to verify that the earth orbits the sun than that the moon orbits the earth. They commit this "scienceblind" error because they still carry in their heads the child's earth-centered view of the solar system.[8] These early views presumably emerge first because they allow tracking of regularities (e.g., the sun "rises" and "sets") with relatively low information-processing demands. Even as I write these words and smugly convince myself that I'd never be earth-centered, I realize the ease of imagining myself lying in a field, with my feet pointing south, watching the sun "move" from left to right across the sky. I find it much harder to imagine the correct account in which I am lying on the surface of a sphere that, when viewed from the North Pole, is rotating counterclockwise such that the sun is revealed in the morning and occluded in the evening by the edge of the rotating earth.

Naive early theories are not always fully displaced by later versions. College students who excel in introductory biology courses can still show influences of childhood essentialism, vitalism, and Lamarckism.[9] In general, early theories are not serious impediments to learning science when instruction addresses them.[10] Two practices may matter most: offering detailed and compelling mechanisms for the correct account and describing and "undoing"[11] the wrong account by explaining its appeal while exposing its critical flaws. When these two practices are done well and the learner continues to employ the replacement, the older view loses much of its influence.

Analogous cases in perception show how we can minimize the influence of first impressions. The moon seems larger on the horizon than in the middle of the sky, but no one actually thinks it increases in size or comes radically closer to us. We know it is an illusion. Similarly, I can talk about the sun rising and setting, and I can even perceptually experience it that way, but I don't really think that's what actually happens. When given time to reflect and acknowledge the illusion's effects, its influence on important decisions becomes negligible. It may only linger in speeded tasks of little consequence. Rising and setting suns, mirages, and cottony clouds do not wreak havoc in our daily lives. We can block visual illusions from creating false beliefs. We can also override early misleading theories by explicitly considering them and viewing them as illusions. We may also benefit by learning how those theories might promote early learning even as they must be set aside with further cognitive development. Associations to the old theories may persist and slow down response times in speeded tasks that require overriding those associations, but when we take care to reflect on the problem or repeatedly give the correct response, the interference is modest.[12]

One way to bring legacies under cognitive control involves the use of related examples that are both familiar and intuitively true. Thus, examples from artificial selection can aid learning about natural selection.[13] Wild mustard plants were guided by humans into different developmental pathways through artificial selection, or selective breeding, into cabbage, broccoli, cauliflower, kale, brussels sprouts, collard greens, and kohlrabi.[14] The human creation of biological diversity can make natural selection more comprehensible. Artificial selection of new species that cannot breed together is rarer than creating within species variation. Very different appearing members of the same species can interbreed (e.g., Newfoundlands and Chihuahuas → Newfhuahuas?), but new species have also been artificially created throughout history.[15] Darwin repeatedly referred to "artificial selection" and domestication, including in his introductory chapter of the *Origins of Species*. Why didn't domestication and artificial selection result in evolutionary theory thousands of years before Darwin? Religious doctrine often blocked the insight by maintaining, falsely, that humans could never create new species and could only tinker with variations with the God-given boundaries of a species.

For several years, the psychologist Deborah Kelemen and colleagues have successfully taught evolution through natural selection to young children through brief lessons.[16] The lessons typically consist of reading special storybooks that provided coherent causal information about critical evolutionary mechanisms such as the links between the following: trait variation in population, food changes in response to climate change, differential health and survival because of differences in getting food, differences in reproduction because of health differences, trait inheritance, and changes in trait distributions over several generations. Children as young as five learned the mechanisms relating these concepts at an abstract level.

That new knowledge enabled them to reason correctly about novel cases of evolution three months later and to resist the pull of their earlier biases, such as essentialism.

New theories, when taught well, can largely replace basic folk misconceptions even in preschoolers.[17] Continued elaboration of the new theory over the course of several years, along with reminders of the hazards of misleading alternatives, may ensure long-term impact. Yet, content continuity and elaboration rarely occur over successive school years. Misleading early theories are rarely discussed in terms of their critical shortcomings, but if wrong cases aren't explained, they run the risk of being strengthened by virtue of simply being mentioned.[18] When we encourage school-age children to ask how and why questions, and more generally, support their natural sense of wonder, we are also helping them enjoy the intrinsic pleasures of surprising new discoveries and deeper insights.

Early sets of beliefs in naive physics, psychology, and biology can influence learning of science and technology, especially when instruction fails to confront the misunderstandings, does not provide sufficiently strong causal conceptual structures to overwhelm earlier models, and stifles wonder. Some aspects of mind, according to Shtulman, are "built to perceive the environment in ways that enhance survival, which do not always align with the categories of science."[19] High-quality, compelling mechanistic alternatives allow us to see these initial ways of thinking for what they really are: useful initial stepping-stones, and nothing more.

Lingering early legacies can facilitate rapid responding. An earth-centered model of the solar system may be horrific for calculating orbits, but it may work better than the real model for quickly guessing "sunset" times. As much as these early models may hinder correct science, some persist because their efficiency provides value.[20] Their value, however, depends on our being aware of their appropriate heuristic uses and of their limitations.

In short, whether misleading legacies are detailed but incorrect models of specific phenomena or biases that skew data collection and representation across all domains, the extent to which they distort everyday thought is heavily influenced by early interventions.[21] Acquired knowledge clusters, also known as "mindware" when they are heavily practiced and automated, can override cognitive legacies.[22] Early legacies may provide us with good footholds for initially making sense of a flood of causal information, but we must wrangle those legacies into appropriate supporting roles, not dominant ones. Guiding children to stronger alternatives is much easier when children's natural drives of exploration and discovery are supported.

Coexistence Dilemmas

Individuals of all ages can believe in two accounts that clash with each other. This problem of *explanatory coexistence* or *explanatory pluralism* is pervasive.[23] We tolerate, and even ignore, blatantly contradictory causal relations. We may happily believe two incompatible relations for decades until the contradiction is explicitly pointed out to us. The psychologist William Brewer has described to me on several occasions how he asks college students to explain why we have summer and winter in places like Illinois. They often reply that the earth is further away from the sun in the winter than in the summer, so it absorbs less of the sun's heat. Brewer then asks what the weather is like in Australia when it is winter in Illinois. People often quickly state that of course it is summer. But then, they often show dawning expressions of surprise and embarrassment. They realize that both beliefs cannot be true, beliefs they have confidently held for decades.

These obviously incompatible internal contradictions may seem rare, but they are commonplace. Think hard about almost any complex system, and you may end up with the realization that two things you have long believed cannot possibly both be true. For example, if I think more deeply about why the highest tides occur at new moons and full moons, I realize I may have contradictory beliefs. I believe a new moon causes extra high tides because the moon and the sun are both lined up on the same side of the earth and summing their gravitational pulls. But there are also extreme tides for full moons, which are on the opposite side of the earth from the sun. I realize that I don't know if the overall gravitational pull of full moons is reduced and creates smaller tides than new moons. If very high tides occurred only once a month during new moons, I would have happily accepted that account. My local tide table indicates that the sun and the moon tugging on the seas from opposite sides has the same effect as tugging from the same side. I can now make up a story as to why, but my original beliefs no longer neatly fit together and I am not confident of my new story. I quickly uncovered similar clashes of causal beliefs when I thought more deeply about how sunflowers follow the sun from east to west during the day, but then realized that they must move from west to east during the night. (I only later learned that sunflowers do not move back at night.)

We rarely have fully worked out explanations of the causal patterns responsible for familiar events. Instead, we have clumps of causal fragments that make some sense on their own. We just assume they'll fit together nicely when we need to flesh out the full account. We don't run into failures very often because we rarely try to self-generate full explanations. Instead, we look them up or ask someone. When we do that, our own a priori fragments may gracefully morph into compatible forms without our realizing the change. When things go well, the fragments help us grasp the explanation more quickly, while not screaming contradiction because we have unwittingly adjusted their early forms to make them fit. If we had been forced to figure out everything on our own, we might have recognized our inconsistencies, or, if a cunning interrogator extracted all our prior beliefs, the inconsistencies would be exposed. This is how gifted trial attorneys are able to discredit even expert witnesses.

Coexistence of conflicting explanations is also tolerated by drawing boundaries between incompatible systems. We make up "rules of engagement" between the two. Both professional scientists and laypeople employ this technique. You can relegate one system as a distal cause and one as a proximal cause. For example, deeply religious evolutionary biologists often say that evolution through natural selection is obviously the basis for the origins of species, but that, ultimately, the entire architecture of chemistry and physics and biology that enables such a process was intentionally designed by God. God then "sat back" and watched it all unfold according to the structures and natural laws that God designed.[24]

A second technique segregates conflicting beliefs to different levels and kinds of explanations. I might believe that causes must temporally precede effects at the macrolevel but not at the quantum level. Less exotically, I might believe that "the more you use x, the better it becomes" and also that "the more you use x, the worse it becomes" by delimiting the first belief to topics within biology and psychology (muscles, problem-solving, etc.) and the second to many physical artifacts (knife blades, sandpaper, racing sails, etc.). We may often set up these boundaries tacitly to reduce conflict. The conflict is especially jarring in David Brin's novel *The Practice Effect*. A planet is described where artifacts don't wear down with use. Instead, they improve. Dull swords get sharper when heavily used, and baskets get stronger the more they are used to carry heavy loads. In addition, artifacts degrade over time if they are not frequently used. Readers initially struggle with such events until they suddenly "get" the practice metaphor and shift all the artifacts into what are normally biological and psychological categories.

A related strategy restricts a belief to a small content domain at the same level of explanation. I may believe that some living things must turn carbon dioxide into oxygen (plants) and that others must turn oxygen into carbon dioxide (animals). Through this sorting of beliefs, we avoid tensions. We just assume that models neatly apply to one domain of things and not to another, even if the real world is messier. We can also switch between systems according to contexts. For example, in competitive contexts one might use zero-sum models to interpret group resource use, whereas in cooperative contexts one might intuitively invoke "positive" sum or win-win models.[25] Similarly, in formal contexts such as a high tea, one might explain behaviors by rules of etiquette, while in informal contexts such as a spontaneous celebration of a sports victory, emotional states may dominate explanations of behaviors.

We can be alarmingly oblivious to how often we contradict the beliefs of our recent selves. An astrology follower may worry about the negative outcomes of an unfortunate alignment of certain stars and planets at breakfast and later in the day pass a physics exam on the inverse square law governing gravitational influence. A surgeon might be superb at sterile technique yet rely on the "five-second rule" when selectively picking up a certain beloved French fry from the floor. These tensions may be the norm for most people with only a few individuals ever bringing a conflict into awareness and addressing them. Moreover, the tensions can range from causal conflicts to lower level associations that don't neatly align. Coral consists of colonies of animals despite strong associations with words for inanimate things ("built on coral," "coral necklace," "coral reef") as well as words for plants ("coral growth" or "coral harvesting"). The most powerful and influential conflicts, however, are those between explanations not just associations.[26]

Surprisingly often, we cope with coexisting contradictory beliefs by doing nothing. We don't actively segregate beliefs by distal versus proximal causation, levels of explanation, content domain, or context. Instead, we passively allow our views to shift in response to local situational factors, such as time pressure or formality. This process often resembles classic accounts of "thinking fast and slow." A rapid imperfect system can easily dominate a slow more reflective one.[27] Speeded tasks can bring to the fore computationally easy heuristics even when we know they are wrong if we think about them at all. When it really matters, however, heavily used and practiced knowledge can come to override fast heuristics even in high-pressure speeded tasks.[28] An experienced rocket designer is unlikely to ever see flames as things, as many people do, and instead always experiences them as complex chemical reactions involving various gases.

A greater metacognitive awareness of the full spectrum of coping processes would enable better monitoring of their possible pitfalls. In addition, well-developed, easily deployed knowledge bundles (theories, models, frameworks) can be flexibly employed to override other initially compelling interpretations.[29] Other times, we may "acquiesce" fully knowing that we are giving into an erroneous easy story but are doing so intentionally.[30] We may give in for reasons ranging from social conformity to risk assessment (e.g., "I think it is highly implausible but there is minimal risk in 'believing' it and a massive cost to not 'believing' it if true"). Santa Claus is one such case for many children of a certain age. More developed senses of wonder promote such metacognitive insights and thereby allow us to more flexibly and strategically manage coexistence tensions.

Illusions of Insight and Understanding—Epistemic Errors

Suppose your toilet isn't flushing well. A plumber says it is because the pipe that directly refills the bowl after the flush is corroded. The copper piping must be replaced all the way back to a major water supply junction in order to remove the obstructing corrosion. The plumber explains this in mechanical detail, and you nod appreciatively because your feel that you are being taken seriously as someone who can understand why the expensive pipe replacement is needed. In this case, however, you'd be a victim of a swindle, aggravated by illusions of understanding. Most people believe they have some sense of how household toilets work while actually knowing very little.[31] In this case, a correct understanding of toilets includes no direct water supply tube to the bowl. But most of us with more minimal understandings will accept the plumber's misleading but right-sounding explanations as resonating with what we falsely believe we already knew.

In addition to legacies' effects on our understandings, our views of knowledge itself matter. These *epistemic* distortions affect beliefs about the nature and amount of knowledge, the acquisition of knowledge, and its emergence in a community. The knowledge that we already have influences what we think we know, often by inflating inferred knowledge far beyond its reach.

One example of knowledge overreach, first described in our lab, is an "illusion of explanatory depth," or IOED.[32] The illusion occurs when people are asked how well they understand how things work. When asked how refrigerators, flush toilets, or zippers work, people tend to confidently predict that they can provide fairly detailed explanations. In fact, they often can only give the simplest caricatures of an explanation. After trying to actually explain, they are surprised at how little they knew.

Some excerpts from interviews in our early studies illustrate the effect:

How does a Refrigerator work?

Participant A: Well, let's see. The electricity makes it cold . . . somehow. Wait. So, there is a motor that makes the inside cold by. . . . Ok, there is Freon in there. [pause] Something to do with the Freon, I think, but how does it get cold? [pause] Damn. I guess I have no idea.

How does a Flush Toilet work?

Participant B: When the lever is pulled, all the water empties to the connecting tank which fills and sanitizes it into connecting sewage pumps. The water refills the tank.

Participant C: You pull the lever, the water in the tank is sucked down, and new water resurfaces.

How does a Zipper work?

Participant D: The zipper is open, then you pull the latch part straight up so that both sides meet. Then you pull the zipper open so that both sides come apart again.

Participant E: The zipper somehow pulls and locks and unlocks the teeth.

How does a Piano make sound?

Participant F: When a key is pressed on a piano, the force of the pressure is connected to the (I don't know what) on the inside of the piano. A note is played as a result.

Participant G: The key acts as a lever—when it is struck, a bar on the inside of the piano is raised. This bar hits another bar, causing it to vibrate and create a sound.

Participant H: You touch the key, and the key in the back of the piano goes down. Then when you release it, the sound.

People who are confident about their explanations have little to offer when they actually have to explain. They merely rephrase the phenomenon to be explained, sometimes trying to use more obscure technical sounding words. In other cases, they make up clearly wrong accounts. Most of the time they quickly realize how badly they overestimated the depth of their knowledge. Occasionally, as they provide a blatantly false account, they start to convince themselves they do understand, but, in such cases, standard follow-up questions reveal their ignorance. For example, several people explained how a helicopter worked by envisioning it as an upward facing rigid propeller. They were using as models wood toys that fly when we spin their vertical posts between our palms. But, when asked how a helicopter goes from hovering to moving forward, they are completely stumped, indeed visibly dumbstruck. They suddenly realize they have no idea of how helicopters actually fly (see figure 7.1).

The IOED has been widely studied across many domains such as biology, physics, political systems, and mental disorders.[33] In our work, it persists in participants who have just acknowledged how badly they overestimated their knowledge of two devices, for example, a cylinder lock and a

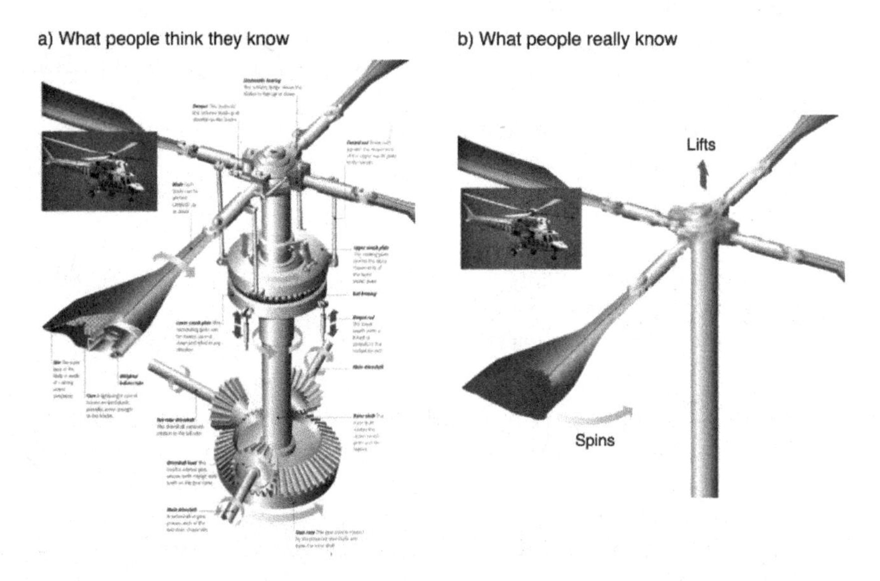

Figure 7.1
Novice and real explanations of helicopter flight. People can grossly overestimate their understanding of how helicopters work. Many adults have not much more than (a) in mind while believing they have something closer to (b).
Source: Keil, F. C. (2003). Folkscience: Coarse interpretations of a complex reality. *Trends in cognitive sciences*, *7*(8), 368–373.

refrigerator. In debriefing, when we explain how other participants were given different items, many participants wistfully remarked that if only they had explained those other items, they would have done much better. In fact, the IOED effects were the same across items.

The IOED is also not just a general overconfidence about ability.[34] The discrepancy between initial ratings of knowledge and later performance is larger for explanatory understanding than for other kinds of knowledge. The illusion of knowing explanations goes far deeper than knowledge of procedures, facts, or narratives. If people are asked if they know how to make an international phone call (a procedure), they tend to be quite accurate in terms of their ability. Similarly, if they are asked if they know the capital of Tasmania (a fact), they rapidly sense that they either know or don't know the answer. The most interesting case concerns narratives, which seem close to explanations. If people are asked what they know about the plot of movies, such as *Titanic*, their assessments are generally fairly accurate.

Several converging processes make the illusion of knowing explanations especially powerful in contrast to other forms of knowledge. First, people often confuse knowing the function of a thing with knowing how it works. I may be adept at using my cell phone GPS but have no idea how geographic position is determined. People have feelings of familiarity and confidence in using things that are confused with having a sense of how they actually work. Sometimes function and mechanisms are so intertwined that fully knowing functions of many components conveys knowledge of mechanism. If a person knows all the functions of the major parts of refrigerator, those functions and their interactions give a sense of mechanism as well, but most people do not have anything close to that degree of layered functional understanding.

The IOED differs from estimating knowledge of facts and many procedures because we can usually quickly self-test whether we know a fact or have frequently performed a procedure successfully. The scaffolding illusion, in which we erroneously think we could have learned a procedure on our own, shows how we can sometimes overestimate our knowledge of procedures, but those cases are limited to carefully structured situations. The IOED is also distinctive because we rarely provide full explanations. We often quickly report facts or demonstrate procedures, but how often do we provide full explanations of how things work? Our greater accuracy for knowledge of narratives may arise because we frequently do supply complex narratives to others. Storylike narratives are powerful ways to increase social bonds; explaining how things work does not normally play such a role.

The IOED is also enhanced by the presence of things we are explaining. We greatly underestimate how much we need the object to be at hand to explain how it works. Ask some friends to explain in detail how a stapler works with no stapler present. They usually soon pause when they can't quite put all the parts together to make a full causal chain. Ask other friends to do the same thing with a stapler in front of them that they are free to manipulate. They will usually give a flawless explanation. They will further believe that a complete understanding was all in their heads even though it deeply depended on having the object present to fill in gaps. The IOED thrives when we neglect the extent to which objects themselves support our knowledge.

We neglect the communities of knowledge in which we are embedded as well. A curious effect, chronicled by the psychologists Steven Sloman and Phil Fernbach, arises from the awareness of communities of experts. That awareness of experts inflates our own sense of understanding as we start to confuse what is in their heads with what is in ours.[35] This makes some intuitive sense. In a way, we *do* know how something works if we can easily access the relevant information in other minds. The philosophers Andy Clark and David Chalmers describe how we offload much of our current world knowledge to devices around us such as cell phones or even pads of paper. They call this the *extended mind*.[36] But we don't offload only to handy objects, we also can offload our knowledge to willing minds of others. Indeed, couples who have been together for decades often do this automatically and seamlessly.[37]

Children have even stronger IOEDs than adults.[38] For children, the IOED may be more adaptive. In particular, equating externally sourced knowledge with what they actually know in their own mind may make them less discouraged by their massive ignorance. They do "know" many explanations by virtue of easy access to well-intentioned caregivers. The philosopher Rob Wilson and I once described the illusion as providing a way of not diving too deep into unnecessary details if they can be easily obtained from other sources when needed.[39] Children with more limited metacognitive skills and weaker working memories may especially benefit from such an illusion as it would help them avoid trying to remember a mass of details and invariably failing. Critically, the IOED does not negate the astonishing ways that young children are able to make sense of the world early on;

instead, it illustrates the importance of abstraction and learning to navigate the terrain of knowledge in other minds. Their relative obliviousness to their degree of outsourcing may help give them illusory but helpful feelings of competence and autonomy. They are, in a sense, getting a feeling of their future competencies.

The IOED is part of a family of related illusions, all of which are stronger in children and which help keep them from being discouraged by their actual ignorance. Kristi Lockhart discussed this idea with respect to the overoptimism found in younger children.[40] All of these illusions lead us to underestimate how much we depend on others to fill out our knowledge gaps, a tendency that can have consequences not only in science but in many other aspects of our private and public lives.[41] We may also underappreciate gaps between ourselves and various expert scientists. Complementary to Socrates's observation in Plato's *Dialogues* that the more he knew, the more he realized how little he knew, it now seems that the less we know, the more we think we know.[42]

Some knowledge illusions involve levels of explanation. For example, when people learn about brain scans that are linked to psychological studies, they tend to believe those studies as being more solid science than when brain scans aren't mentioned. Studies in our lab showed that people easily reject as empty and circular short summaries of research findings without neuroscience, but they embrace those same circularities as important and insightful when they are linked to uninformative information about brain scans.[43] The mere mention of neural measures converts scientific skeptics into gullible victims. More broadly, people tend to put too much weight on more "basic" sciences in explanations, revealing a reductionist bias. Children exhibit that tendency to an even greater extent, grossly underappreciating, for example, the complexity of psychological phenomena such as the ability to learn a language.[44] More extreme cases remind us of the danger of impulsively embracing lower levels. The philosopher Jerry Fodor once described the folly of trying to explain the principles of economics in terms of the physical motions of currencies,[45] a task that seems particularly silly in the era of digital blockchain currencies. In such instances, we can easily see the implausibility of developing coherent, comprehensible mechanistic accounts that smoothly work across so many levels.

A final knowledge illusion occurs when we use search engines. We assume that some of the information that we can easily access is in our heads. To explore this idea experimentally, we asked some people to search for how a toilet works and asked others to read the same description on a piece of paper (the search was rigged to give the same result as the paper version). The search-enabled people thought they knew more about all sorts of other searchable information (e.g., "How well can you explain how a refrigerator works?") but *not* about information that is obviously not searchable (e.g., "How well can you explain why your father choose his career?").[46] We easily confuse ease of access to information with knowing it on our own.

These studies came to mind because of a peculiarity of our home in Connecticut. We live on a peninsula that sticks out into the Long Island Sound. Our house is also in a location where there is no cell phone service. When a major hurricane comes through, the access road becomes flooded with several feet of water and all electrical power stops. We become completely unplugged from the rest of world. This has happened twice in the twenty-three years my family has lived in Connecticut, and on both occasions, I felt myself becoming dumber by the hour. I had forgotten how often I searched for information on my laptop or cell phone. When suddenly deprived of this resource, I felt increasingly ignorant. This particular illusion of knowing is clearly linked to the rise of the internet and search engines. Forty years ago, when we lived on a lake in Upstate New York, we also were completely cut off for days at a time by snowstorms. But we didn't have the same sense of knowledge loss. However, we did have an extensive collection of reference books, including the 1929 edition of the *Encyclopedia Britannica* and a large bookshelf of more recent almanacs, atlases, and science books. We turned to this collection of books many times a day. There may have been an illusion of knowing induced by the presence of the bookshelf nearby. However, it was much less than with search engines. The ease and immediacy of access to information provided by modern search engines makes our dependence on the internet much less obvious than when we thumb through books. In that sense, books may have instilled a greater degree of intellectual humility.

Well-maintained wondering and sustained immersion in the PHED cycle help manage illusions of knowing. We can see our gaps in understanding by being more attuned to puzzling phenomena and asking the right sorts of questions. We can focus on mechanistic understanding in particular to get clarity on what we do and do not know. We can better appreciate the divisions of cognitive labor around us, how they can foster illusions, and why that sense of knowing can serve a useful purpose. We see how to be cognitively comfortable with our gaps rather than complacent with misleading feelings of knowing.

Polymaths: Wandering Wonderlust?

Although ordinary mortals can override early erroneous theories and knowledge illusions through carefully delivered mechanistically accurate details, how reasonable is it to assume that most of us could do so frequently on a daily basis? Perhaps the ability to constantly resist the sirens of misinformation is too much for most of us. Given the massive causal complexities in the natural and engineered worlds, complexities that increase with each new discovery or invention, maybe we can't transcend the detours that we inherit from our childhoods. Ironically, those rare and unusually gifted people who do seem to "know everything" show how all of us better manage our mistaken beliefs and knowledge illusions. Such polymaths or "Renaissance" individuals repeatedly argue that they only differ from us because their early experiences of wondering became so rewarding and mind-expanding that they couldn't help but continue to wonder.

What happens when an energetic child becomes fully addicted to wonder? One supposed outcome, at least until the early nineteenth century, was a universal knower, a Renaissance person, a polymath. Given the growing complexity of science and technology, today a true polymath with full mastery of the critical ideas across all the disciplines seems preposterous. Was it ever possible? Many have described Leonardo da Vinci as a polymath given his achievements in architecture, anatomy and physiology, botany, physics, engineering, and, of course, painting (among many others). One biographer, Walter Isaacson, in a passage that very much supports the propulsive force of wonder, wrote, "The reason he wanted to know was because he was Leonardo: curious, passionate, and always filled with wonder."[47] After Leonardo, a large cluster of acclaimed polymaths appears between around 1700 and 1810: These include Gottfried Leibniz, Isaac Newton, Benjamin Franklin, Mikhail Lomonosov, Thomas Jefferson, Mary Somerville, and Erasmus Darwin (Charles Darwin's grandfather).

No one actually thought these individuals had exhaustive knowledge of absolutely everything. For many crafts, such as weaving, it may take decades to become an expert. Before the GPS age, London taxi drivers might take decades to fully learn the geography of that city. No single person could possibly achieve full mastery in all areas at any point in written history. Instead, the ideal polymath in 1800 knew everything that mattered in all realms of study, broadly defined. They had deep foundational understandings of all major topics in nature and technology as well as major strengths in the arts and humanities. They knew classes of mechanisms, causal patterns, and abstract formal principles, not trivia or mere facts. They could immediately contribute in major ways to the most advanced scholarship in any discipline. A polymath would be expected to know all the important principles and mechanisms explaining animal and plant functions, how all the major classes of engines and machines worked, how the planets and stars moved, and about how weather happened. The polymath should know about chemistry, botany, and all areas of mathematics and logic. In addition, a sampling of the humanities and the social sciences was expected but rarely a comprehensive knowledge. There seemed to be a tacit assumption that full coverage in those areas was unnecessary; just a few spectacular demonstrations would suffice. Unfortunately, the proposed polymaths were almost always men.

In the year 1800, such a comprehensive understanding of everything did not seem obviously impossible. The full knowledge of how the human body worked was a tiny portion of modern medicine and biology. Chemistry was just beginning to take off, and the rapid growth of technology was in its first stages. We may not realize today the full limits of science understanding a little over two hundred years ago. The science writer Edward Dolnick provides a sense of these limits in his descriptions of topics of serious debates at the Royal Society (the leading scientific society in the world at that time) in around 1700.[48] The society experimentally tested the view that a spider placed in the middle of a circle composed of the powdered horn of a unicorn would be unable to leave the circle. It was fortunately disproved. An eminent chemist in the society was apparently convinced that cataracts could be cured by blowing a powder made from human feces on to the afflicted eyes. Several leading members of the society were convinced of the existence of mermaids.

The topics became less exotic by 1800 but still included titles such as "Account of a Monstrous Lamb," "Account of an Elephant's Tusk, in Which the Iron Head of a Spear Was Found Imbedded," "Observations Tending to Shew that the Wolf, Jackal, and Dog, Are All of the Same Species" (with claims of successful progeny), and "Some Observations on the Irritability of Vegetables" (with proposals of shared mechanisms with irritability in animals). It is a remarkable experience to read through the Proceedings of the Society for the eighteenth century. You first realize that the vast majority of entries, apart from several entries in mathematics, would be easy to understand by people from diverse backgrounds. You also see an extraordinary spread of topics, ranging from revolutionary and beautifully executed studies in astronomy, chemistry, and physics to more fantastical and bizarre entries. As the century ends, there is an almost palpable feel of knowledge accelerating at a pace that will soon surpass the capacity of any one person. There is also the sense of science emerging from a morass of other more dubious topics.

With sufficiently abstract explanatory and mechanistic knowledge and no requirement of all the trivial details, some individuals did make major contributions in an astonishing range of areas. They were not all rich lords who didn't have to work, and they were not all provided with the best possible educations. Then, at some point, in the early 1800s, this kind of polymath disappeared as science and technology flourished. At least five different books begin with the title "*The Last Man Who Knew Everything*." They select different people, but they all assumed that knowledge progressed to a point where polymaths who truly knew everything important were no longer possible.

My favorite polymath in this first sense is Thomas Young (1773–1829). Young is not as well-known as some others but is arguably the most impressive polymath of all. From 1802 to 1803, he gave a series of almost a hundred lectures at the recently founded Royal Institution and at the older Royal Society. These lectures covered virtually all the known sciences and much more. He proposed the wave theory of light and highlighted critical flaws in Newton's "corpuscular" approach. He formulated Young's modulus of elasticity, a foundational concept in engineering. He discovered several basic mechanisms of human vision, including accommodation, astigmatism, and the three-component theory of color perception. He came up with a key insight that led to the later deciphering of the Rosetta Stone by Champollion. He derived equations governing capillary pressure and surface tension. He made several discoveries about the physiology of the heart and circulatory system. He was familiar with a vast number of languages and compiled the authoritative taxonomy of over 400 languages based on their grammars and etymologies. He was an expert on Egyptology. He invented a way to tune musical instruments as well as developing a theory of musical harmony. He devised a way to calculate life insurance, developed models of ocean tides, and proposed methods of carpentry—and this hardly exhausts the list[49] (see figure 7.2).

How did Young arrive at such extraordinary depth and breadth of knowledge? He attributed it to a lifelong drive to learn, and he often referred to the pleasures of learning and achieving insight. He said that working on challenging intellectual problems is what kept him alive. Indeed, on his deathbed, as a friend urged him to stop working on an Egyptian dictionary for fear of exhaustion, he replied that he very much wanted to continue because it was a source of "great amusement."[50] Throughout his writings, an ever-present wonder seems to impel him to learn more and more about all possible things. He came from a middle-class family of ten children and was financially secure, but he was not especially wealthy or well educated by any kind of formal instruction. His drive to learn, explore, and discover seemed much like a preschoolers' wonder on cognitive steroids.

One of Young's most remarkable insights concerned his views of the cognitive processes involved in building useful concepts about so many different things. He wrote,

> It seems to be wise and benevolent . . . that a certain degree of oblivion becomes a most useful instrument in the advancement of human knowledge, enabling us readily to look back on the prominent features only of various objects and occurrences, and to class them and reason upon them, by the help of this involuntary kind of abstraction and generalization, with incomparably greater facility than we could do, if we retained the whole detail of what had been once but slightly impressed on our minds.[51]

Here, Young neatly summarizes the essence of what children and adults alike must do to cope with the overwhelming complexity and details of what they encounter every day. They must form abstractions out of experienced details and then leave the details behind. Moreover, these abstractions must enable the ability to generalize to a wide range of new cases. Along with everything else, Young was a terrific cognitive scientist! He also anticipated how abstraction and generalization would be central to cognitively coping with ever-expanding scientific and technological knowledge.

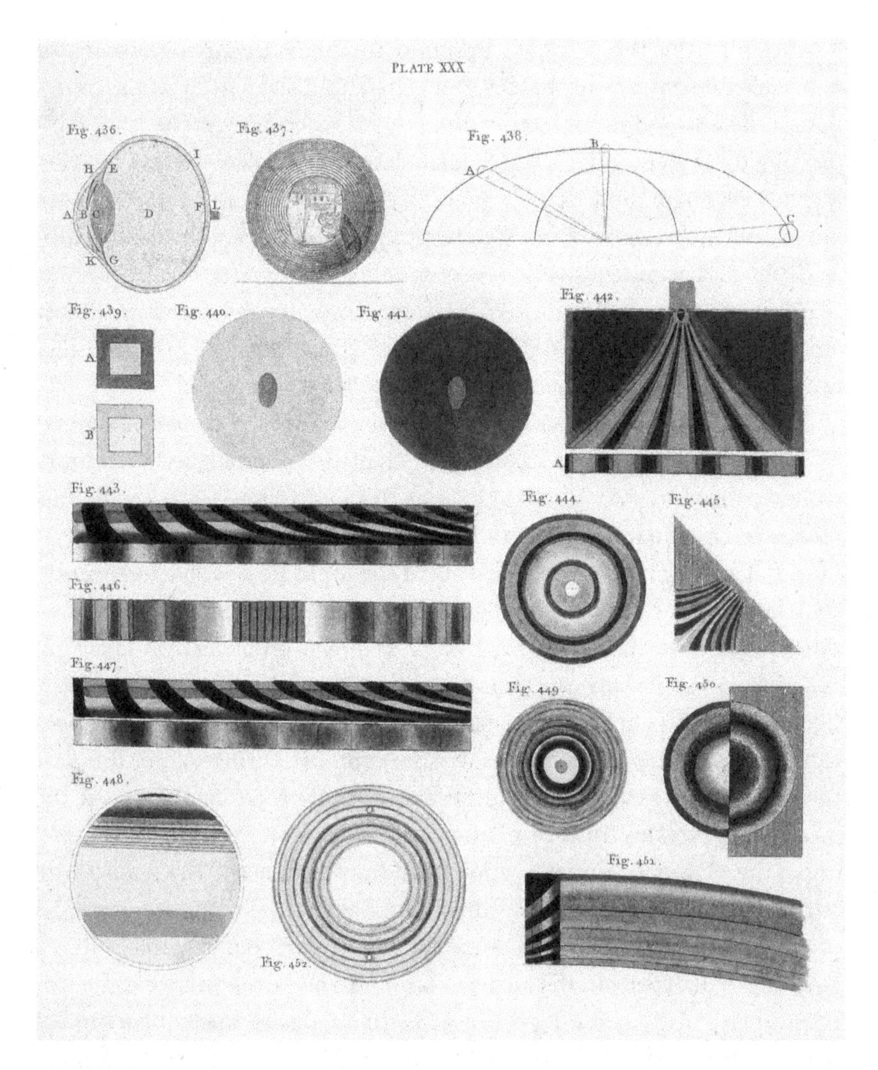

Figure 7.2

This set of drawings, is one just one page from Thomas Young's 1802 lectures to London's Royal Institution. It illustrates, among other topics, Young's knowledge of eye structures, his description of image inversion on the retina, and his double slit study of light (which led to the wave theory of light and ultimately was foundational to quantum physics). It just one of 63 illustrated pages full of important discoveries across the sciences and engineering. https://en.wikipedia.org/wiki/Thomas_Young_(scientist)#/media/File:Young-Thomas-Lectures1807-Plate_XXX.jpg.

After the early 1800s, there was simply too much for any one person to know even if someone abstracted and generalized away from all the unimportant details. A different sense of polymath emerged, with its last exemplars in the 1960s and 1970s. At that time, people like Norbert Wiener, Richard Feynman, and Herbert Simon were notable examples. They were polymaths in the sense of being universal problem solvers and rapid learners. They had acquired diverse sets of foundational models and impressive cognitive tool kits that enabled them to rapidly come up to speed in almost any area of inquiry. They didn't know everything, but they could quickly learn anything by building on their prior explanatory systems.

Stories abound of Wiener's ability to enter almost any conversation and to be soon offering novel solutions to challenging problems, sometimes while engaged in two conversations at the same time. Herb Simon was interested in virtually anything that presented an intellectual challenge. If he didn't already know a great deal about the topic, he was able to instantly see relations to areas where he did have well-developed understandings.

Feynman was able to quickly diagnose novel problems. He famously determined the precise cause of the explosion of the space shuttle *Challenger*. The O rings in the booster engines had failed to expand at 32 degrees Fahrenheit and thereby let explosive gases combust. He discovered this flaw through careful observation and questioning. He then demonstrated the problem by dunking an O ring into a glass of icy water at a televised meeting of the national commission investigating the disaster. The science and engineering aspects of the problem were not at all close to his primary areas of expertise in physics, but in a few days, he figured out a problem that a large group of experts in the area had been puzzling over for months. Feynman is particularly revealing because he often reflected upon the nature of scientific thought. The following excerpts from interviews and Feynman's writings[52] summarize many of the central themes of this book:

> "I was an ordinary person who studied hard. There are no miracle people. It happens they get interested in this thing and they learn all this stuff, but they're just people."

> "I don't know anything, but I do know that everything is interesting if you go into it deeply enough."

> "I learned very early the difference between knowing the name of something and knowing something."

"The third way to tell whether our ideas are right is relatively crude but probably the most powerful of them all. That is, by rough approximation. While we may not be able to tell why [Alexander] Alekhine moves this particular [chess] piece, perhaps we can roughly understand that he is gathering his pieces around the king to protect it, more or less, since that is the sensible thing to do in the circumstances. In the same way, we can often understand nature, more or less, without being able to see what every little piece is doing, in terms of our understanding of the game."

"We absolutely must leave room for doubt or there is no progress and there is no learning. There is no learning without having to pose a question. And a question requires doubt."

". . . the scientific spirit of adventure—the adventure into the unknown, an unknown that must be recognized as unknown in order to be explored . . ."

"I am always looking, like a child, for wonders I know I am going to find. Maybe not every time, but every once in a while."

These excerpts are especially revealing in the ways they appeal to cognitive abilities that we all share. Feynman stresses ways of thinking and learning that are present in every child and that we should retain and further develop. Much of the time they also resemble different aspects of wonder.

While it was no longer possible to know everything in the 1960s and 1970s, it was still possible for a few extraordinary people to quickly learn about diverse new areas and engage in discussions and problem-solving at the highest levels. They couldn't stay on top of everything at the same time but were able to dive deep into what was known about virtually any system and soon function at the same level as leading experts.[53]

Now, in the 2020s, that second sense of polymath seems to have also disappeared. Many mechanistic accounts in the sciences and technology are so complex that it can take years to be a full participant in the highest levels of discussion. There is simply too much background knowledge required. Remember that my 1963 Triumph Spitfire had no transistors while today's cars have billions. Circuits are embedded within circuits in nested arrays that may be hundreds of layers deep. Modern molecular biology must frequently rely on massive data analytic and computational techniques to make sense of gene pathways and their interactions. Researchers are ever more specialized, and the number of authors on single papers can run into the hundreds. Between 2012 and 2018 the number of papers with over a thousand authors went from none to a hundred.[54] More broadly, the average number of authors per paper in the sciences has increased by more than a factor of five in the past century.[55]

A third, and perhaps final, sense of polymath may be coalescing today in the 2020s. Those individuals don't know everything and can't quickly come up to steam to be equal to leading experts in most areas, but they are gifted at knowing how to identify the leading experts and at knowing how to extract usable explanations and insights from them. Such individuals still must know a great deal about abstract causal patterns and mechanisms across the sciences and technology to know how to locate experts, but they can't achieve high levels of real-time expertise and be true peers. Like the other two kinds of polymaths, however, these individuals may also be those whose passion for wonder flourished continuously from early childhood on.

Who is in this third groups? They may be an evolved version of the *mavens* described by Malcom Gladwell in *The Tipping Point*. Mavens were individuals who simply "knew who knew." If you wanted an expert opinion on something you'd go to a maven. We all were familiar with such people back then at the beginning of the twenty-first century. They were plugged into information in ways that made them uniquely helpful resources. They were human search engines. No one talks about mavens anymore because we all use Google or other software to query virtually anything. But only a handful of individuals may have the uncanny ability to sniff out the best sources of information on any problem. They work through some combination of internet searching, consumption of other online media, and face-to-face conversations, and, most importantly, they have a strong foundation of knowledge about recurring causal patterns and mechanisms across diverse topic areas.

Almost every week, a discouraged undergraduate tells me they can't find any articles on a topic he or she would like to write about. After a minute or two of clarifying what that topic is, I am almost always able to refer the student to at least a dozen or so papers that the student is startled to find extremely relevant. When asked how I found those papers, I shrug because it is largely an intuitive process built up over many years of searching for information in psychology and the cognitive sciences. Today's polymaths can do this for any topic and can formulate questions to leading experts that those experts would regard as central to the most active areas of research in their discipline. Those people are probably relatively young, closer to the generation of "digital natives." They clearly have a sense of the digital lay of the land that few, if any, in my generation have. They constantly track many streams of information and monitor how their usefulness waxes and wanes over time. But they also have learned much about how things work at abstract levels and use that knowledge continuously to explore and evaluate potential sources of information. Those future polymaths may not be university professors and leading scholars. The center of gravity for this new breed may be shifting away from the university to other sorts of real-world and online communities.

Is the "mechanism desert" making polymaths impossible by burying so much of the causal structure in electronic circuits? In the third sense of polymath, one that employs more skeletal understandings and a sense of the social distributions of knowledge, we can still become more like those Renaissance minds than we ever imagined. To get a better sense of how this can happen, we once again must look to our younger selves. With appropriate support and encouragement, children can encounter rich and diverse arrays of vivid, real-life mechanisms. These provide the means to build those explanatory tool kits enabling them to engage meaningfully with almost any area of inquiry. We don't need to know everything; we instead need a sense of the right kinds of questions and answers. We won't all become true polymaths even in this third sense, but it is at least possible to see how sustained and encouraged wonder could enable us to come much closer than most of us are now.

▪ ▪ ▪

Developmental legacies can become cognitive handicaps in adulthood. Convenient fictions in childhood can spawn inconvenient fictions later on. Early core theories and shortcuts that get us off the ground may bring us crashing to earth as adults if we don't acknowledge them and manage them appropriately. In the best cases, they serve as healthy early footholds and survive as useful shortcuts in special situations. They are best managed by cultures and caregivers that carefully nourish inquiry and explanation building throughout childhood. We can counter cognitive legacies and detours by building up explanatory schemas and cognitive interpretations that provide more plausible and compelling explanations. A lifetime of nurtured wondering can help build robust senses of abstract and generalizable causal patterns that can serve as building blocks of sturdy explanations. Knowing how classes of things typically work is a powerful antidote against subjectively compelling first impressions. When such knowledge about the world is combined with metacognitive knowledge about our biases, cognitive legacies lose their seductive power and can play useful supporting roles.

The IOED and related illusions create false senses of cognitive satiation when we are unaware of them. Well-developed wondering, however, brings the illusions into awareness and blocks their negative effects. When you wonder about different possible explanations, you come to see current gaps in understanding, the intrinsic uncertainties with any explanation, and the possibilities of future conceptual revisions, and even revolutions. Wondering provides a broader epistemic perspective and engenders greater intellectual humility. It may also help us sense when illusions of understandings may be useful. Feelings of knowing may signal the ease with which we can gain deeper understanding through accessing other minds, accessing the internet, or having the object at hand. We can learn to identify when that feeling of knowing is reflecting a form of indirect knowing and not merely an empty illusion. But when we don't engage in this way, and are devoid of all wonder, we become susceptible to far more than childhood biases and illusions of knowing. Our minds become petri dishes for growing epidemics of misinformation that lead us not just to mistaken beliefs but also to denial and distrust of the best that science has to offer.

8 It's a Wonderless Life

Several years ago, I was on flight from New York to San Francisco sitting next to a balding fellow passenger in a rumpled suit. He was very friendly. We soon discovered that we both had gone to MIT, he several years ahead of me. He worked for a large multinational company as head engineer in a division that manufactured jet engines for passenger and military aircraft. He loved his job, and he loved jet engines. Shortly after taking off, he asked me what I thought about climate change. I said I believed that temperatures were rising largely because of increased human production of carbon dioxide. I said the increase had created a greenhouse effect in which visible light was let in through the atmosphere, but infrared wavelengths were reflected back to earth by the CO_2, thus trapping the heat. My seatmate agreed that CO_2 could trap infrared light, but then asked how much carbon dioxide I thought was in the atmosphere anyway. I said I had no real idea but I assumed as must be at least 10 percent and probably closer to 30 percent of all the gases given the huge impact it had. He grinned and said it was a bit less—it was about 0.04 percent or 400 parts per million.

He seemed sincere and knowledgeable, and I decided to believe him at least for the moment. But I was confused. How could such a relatively small amount have such a huge effect? He asked me whether I thought that carbon dioxide was the largest driver of temperature changes. I said it surely was, but he replied that in fact water vapor causes 60 percent of the greenhouse effect and was massively more present in the atmosphere, by a factor of roughly 100. He added that lowly methane was 25 times more effective in trapping heat and that nitrous oxides (often found in smog) were 300 times more heat trapping than CO_2. Given that CO_2 was such a tiny fraction of the atmosphere and given that it wasn't close to being the main driver of global warming, why was I so convinced that human-produced CO_2 was causing climate change? He added that, even worse, CO_2 was only a small part of other atmospheric CO_2 produced by natural means.

I said there must be experimental proofs of human-produced CO_2's effect on warming. He agreed that a good properly scaled experiment would be great but said that virtually all evidence was correlational accompanied by tiny toy experiments and cumbersome computational models that were difficult to interpret. I asked him why so many scientists believed in human-caused global warming. He said that people tend to believe what they want to believe because of their peer groups and irrational fears and desires. The negative consequences of excessive warming are so scary that they feel morally bound to believe just to be safe. He added that most scientists were liberal and, as such, were deeply distrustful of large corporations such as his own that made jet engines.

My flight companion agreed that global warming probably was occurring but asserted that there were much more reasonable causes than increases in human-made CO_2, given its miniscule amount. Possible causes could range from sunspot activity, to unusual jet stream deviations, to shifts in algae reflectivity in the oceans. He reminded me that massive temperature changes occurred in the ancient past without any human activity, namely the great ice ages, and said it was still unclear exactly what started and ended them. Why should we be so confident that tiny amounts of human-produced CO_2 were the cause today, especially since human-produced CO_2 was only a small part of all CO_2? The conversation went on at this level of detail for hours, covering everything from the flaws with various computational models, to mitigating effects of clouds, to lagged regression analyses and equations for nonlinear dynamic systems. I was constantly off balance and unable to coherently respond to his arguments.

I got off the plane in California in a state of cognitive disarray. The instant I arrived at my hotel, I frantically Googled everything I could about climate change in a desperate attempt to reassure myself that I wasn't part of a vast left-wing delusion. It took a while to regain my footing. I first learned that the numbers my fellow passenger cited seemed to be correct. He wasn't fudging the data in any obvious way. But I then learned how tiny amounts of particulate matter and oxide molecules could produce intense smog that occluded entire mountain ranges. Ozone, the major component of smog, is a far smaller percentage of the atmosphere than CO_2's tiny 0.04

percent, yet, smog can be highly visible, toxic, and physically irritating. Miniscule percentages of tiny parts can have massive effects, and CO_2 is one such case despite its status as a "trace gas." CO_2's molecular structure absorbs infrared wavelengths at much higher levels than more common atmospheric gases and thereby keeps more heat from escaping to space. I learned that climate change is complex in special ways that make it an outlier for easy public understanding. Full-scale experiments with control conditions are not possible. Instead, bits of experimental data are combined with huge computational models of a highly dynamic system, with many experts squabbling about the details.

When I returned to Yale a week later, I ran into an eminent earth scientist and asked him how I could get a clearer understanding of the definitive proofs that global warming was largely due to human activity. He shook his head and said that even he had to take lots of components on faith. He put a lot of weight on the unique isotope signature of CO_2 produced by fossil fuels and how that kind of CO_2 so closely correlated with temperature rises, but the statistics were correlational and there were time lags and other irregularities. My colleague said it was nothing like discovering a new virus to explain a disease, or even like measuring how many neutrinos pass all the way through the earth each second. It was a very different kind of problem. Indeed, in a *New York Times* interview on June 12, 2019, the Oxford geophysicist Tim Palmer sharply criticized state-of-the-art computational models in terms of their predictive accuracy.[1] He said they included questionable assumptions about how to scale up models of turbulences and eddies and, because of limitations on computer power, made computations using much too large cubes or "grid cells" of the atmosphere.

Controversies over climate change show how neglecting wonder, suppressing questioning, and losing intellectual humility can lead to systems of misleading beliefs. Construction of these elaborate accounts can be self-reinforcing. They deploy convincing mechanistic fragments buttressed by cognitive detours and illusions of understanding. To defend ourselves from such threats, must we be constantly hypervigilant and have a band of enlightened experts constantly always on call? In most cases, less onerous defenses will suffice. If our early tendencies to wonder are allowed to flourish and become ever more powerful, far fewer people would construct such elaborate distortions of reality. Critically, when we do encounter such distortions, we'd soon see through such knowledge charades. A growing system of wonder begets a wisdom to discern explanations that respect reality.

Throughout this chapter, we'll encounter egregious cognitive and social consequences of neglecting wonder, especially in cases in which complicated causal patterns and motivational factors are at work. We'll look at how distrust, disengagement, and denial (the "bad Ds") can proliferate when we stop asking why and how and cease contemplating different possible explanations. When wonder withers, we are much less likely to develop our abilities to navigate the divisions of cognitive labor. We also lose the insights provided by mechanistic explanations and the ways abstractions from specific explanations help us interpret and evaluate new explanations in real time. But we will also find encouragement by seeing how a life full of wonder cultivates deft use of doubt, deliberation, and deference (the "good Ds") and enables us to better track truth.

Climate Change and Cognition

For climate change, the total body of information about the relevant scientific communities, their practices, and the feasibility of proposed mechanisms makes an overwhelmingly strong anthropogenic case. Massive amounts of respected science strongly indicate that humans are largely responsible for global warming in the last century. But even the strongest empirical findings are not "proven facts." The broader public needs to understand the difference. Revisions of theories and predictions are intrinsic parts of science and should not be seen as fatal flaws. In addition, most of the public who believe in human causes of global warming do so as leaps of faith. Blind faith makes them fragile knowledge stakeholders who are easily swayed by motivational pushes and disinformation campaigns. Most people have minimal mechanistic knowledge of the phenomena in question, even when they endorse the scientific consensus. One study summarized the public's mechanistic understanding of global warming with the following statistics: Only 12 percent had a vague sense of gases trapping heat. Only 3 percent actually discussed the greenhouse effect, and no one mentioned the critical idea of differences between infrared energy and visible light in terms of trapping greenhouse gases. Yet a large majority of the same sample thought global warming was human caused.[2]

While the greenhouse metaphor is relatively simple, fossil-generated CO_2's effects are harder to encapsulate in a mental model. Correlational data, computational modeling, and statistical inference create cognitive challenges for most people. The diversity of relevant data overwhelms attention: ocean currents, plate tectonics, warming over land masses, polar ice caps, exchanges of water between many different reservoirs (the oceans, atmosphere, land surface, biosphere, soils, and groundwater systems, and rocks), different types of greenhouse gases and how they are produced, cloud cover patterns, algae growth in the oceans, and many more. Difficulties in thinking about causal patterns over long time intervals, another legacy from childhood, poses yet more challenges.[3] People often confuse weather, day-to-day fluctuations in temperature, with climate, changes occurring over decades and even centuries. Finally, climate change is about stochastic processes, not precise deterministic chains. Climate scientists can inadvertently give deniers ammunition. More than 97 percent endorse human-produced CO_2 as a major driver of global warming.[4] Yet, for fear of undercutting this message, they can gloss over the intrinsic uncertainty of their models.[5] Their understandable reluctance to disclose such limitations enables others to misleadingly suggest the models are useless, when in fact they are completely convincing once the limits of their precision are understood.

Climate change may involve especially opaque mechanisms and unusually complex statistical models. Scientists who specialize in climate research cannot easily offer explanations that go beyond the greenhouse idea, the signature of fossil fuel–generated CO_2, and a few correlations. To my knowledge, the computer modeling research has never been presented in any detail to the general public. For all these reasons, the distorting effects of community motivations, consensus, local cultures, and motivated reasoning may be especially strong. A decade ago, the story seemed especially bleak when my Yale Law School colleague Dan Kahan showed that cultural consensus and motivated reasoning could keep climate change denialists from changing their minds. Some scientifically sophisticated denialists were even able to weaponize science fragments and use them to increase doubt among less sophisticated believers.[6]

Despite these gloomy results, later studies showed that compelling scientific explanations can move climate change attitudes toward the scientific consensus. The most successful transformations of attitudes have focused on greenhouse effect mechanisms. People read passages explaining how the earth transforms visible light into infrared energy that is then trapped from leaving earth by atmospheric greenhouse gases. The passages also showed how human production of those gases has been tightly linked to trapping more and more infrared energy. That well-crafted message changed deniers into believers and strengthened weak believers into strong ones, creating new attitudes that endured for extended periods.[7] Mechanistic explanations not only transform beliefs about climate change, but they also increase intellectual humility by reducing illusions of understanding.[8] When we build on our early mechanistic mindsets, we are better able to use new information about mechanisms to adjust our views.

We can also inoculate the public against climate change misinformation.[9] First, scientific consensus must be communicated and justified. This is then linked to warnings about agents who are motivated by politics and/or economics to create doubt about scientific evidence and models. Finally, we can describe the likely content of disinformation attacks and why they are incorrect fabrications. These inoculations are labor-intensive and require considerable advance intelligence about the attacker. They need to be specific as well. General science literacy may not provide a strong enough shield. It must be supported by skills inherent in sophisticated forms of wondering, namely knowing how to doubt, deliberate, and defer.

Is climate change atypical? Do other cases of pseudoscience occur in the same way as climate change denialism? Comparable instances wherein social consensus and cultural norms override mechanistic knowledge are uncommon in Western cultures. Thus, no overriding effects are found for views of genetically modified food or gene therapy when compared to climate change.[10] For genetically modified food and gene therapy, detailed mechanistic knowledge reliably predicts greater rejection of pseudoscience. Unfortunately, the most extreme opponents of genetically modified food think they know the most about the topic when they typically know the least.[11] This makes them less likely to seek out further mechanistic information and more likely to actively shut out mechanistic information as unnecessary. They may also be those unfortunate people whose preschool inquisitive, exploratory natures were the most trampled in later childhood.

Biology beyond Mechanisms

Evolutionary theory is another case that relies on far more than clockwork mechanisms. In addition to being distorted by early core theories and biases,[12] evolutionary theory poses new cognitive challenges. For example, humans do not easily grasp temporal patterns that unfold over thousands and millions of years. We expect most events to occur in a manner of minutes or hours. People are easily so confused by the timescales of evolution that they are often surprised to learn that dinosaurs roamed the earth for far longer periods than the existence of humans. Probabilities and probabilistic distributions are critical concepts. Tigers seem to be one kind of thing with a fixed essence. Instead they are a distribution of phenotypes and genotypes with no one genetic sequence shared by every tiger. Without such distributions, natural selection could not operate. Species must vary in their fitness for different environments. Children, and many adults, do not find such ideas easy to grasp. Evolution itself is a probabilistic process that must be described over an entire population, not just individuals. Finally, evolution is difficult to visualize. It would be much easier if individual animals slowly morphed into other species in a single lifetime, much as tadpoles turn into frogs and caterpillars into butterflies. Instead, one must think about large groups of animals over many generations and how environments shift distributions of phenotypes by influencing distributions of genotypes. Any area of science that shares some of these properties may present special cognitive challenges. One provocative class of examples at the intersection of biology and chemistry concerns toxins and carcinogens.

In 2018 a groundskeeper in California who was dying of cancer was awarded $289 million in a lawsuit against Monsanto, the maker of an herbicide known as Roundup. The jury was asked to consider the scientific evidence as well as alleged attempts by Monsanto to cover up the risks of Roundup. The case was appealed and the award somewhat reduced, but many other cases subsequently appeared with total claims running into billions of dollars. The case is complicated by largely irrelevant, but unpleasant, strategies used by Monsanto to try to dominate various agricultural sectors. With respect to explanations, carcinogens can be especially difficult to understand in simple causal terms.

Many people who heavily use Roundup don't get cancer, and those who do get very different forms of cancer. People get cancers for many reasons, and a large number of people also use Roundup. Intersections of the two populations are inevitable and do not necessarily prove a causal connection. Considerable sophistication with statistical analyses is needed to evaluate such claims. Virtually all experts without conflicts of interest agree there is no quality statistical evidence that the key chemical in Roundup, glyphosate, causes cancer.[13] Yet, several court cases have resulted in massive awards.

Given the well-known distorting effects of salient individual tragedies in lawsuits and political campaigns, juries who are befuddled by the science can tilt toward the plaintiff bringing the suit. It is hard to think in terms of population statistics when confronted with a suffering individual. Judgments are further distorted when jurors have unrelated reasons for negative attitudes toward Monsanto, Roundup, or herbicides in general.[14] In terms of possible mechanisms for carcinogens, the accounts are maddeningly complex and controversial. They require knowledge of organic and inorganic chemistry and of a host of direct and indirect interactions with DNA and often involve extended debates on the credibility of current classification schemes and models.[15] A recent review discussed eighty-six different mechanisms.[16] Such mechanistically challenging cases can be easily distorted in cognitively and emotionally compelling ways that go against the science.

When accounts become as tangled and complex as the Roundup case, our abilities to evaluate them on our own are limited. A focus on mechanisms may not help us if so many different ones are involved. Yet, we can still ask for explanations of why the evidence is being interpreted as strongly supporting the causal claim, one form of doubting. We can ask how the strength of the evidence compares to that for smoking as a cause of cancer, deliberating the details. We may make the most progress, however, when we consult the right experts and defer skillfully. Given that even preschoolers have early skills at evaluating sources of knowledge, if we continue to grow those skills through high school and beyond, we will be better able to sense which experts are more trustworthy.

Continued Influence of Early Intuitive Theories and Biases

Cognitive vestiges from childhood can aggravate the effects of disinformation when they run unchecked by a lack of questioning and explanation seeking. Early distortions, such as those concerning evolution, may endure most strongly when they align smoothly with cultural consensus and motivated reasoning. To shake the hold of a vestige such as essentialism, educational programs can override legacy-driven pseudoscience such as creationism, even in young schoolchildren.[17] Few researchers, however, have attempted comparable educational interventions with believers in a flat earth, astrology, or telepathy, all of which may exploit vestiges. Because the numbers are small, we tend to see such groups as curious, and even amusing, rather than threatening. More worrying are cases in which most people embrace clearly flawed science, often at considerable personal expense. The most problematic versions occur when a convergence of several early legacies create especially robust false beliefs, even though those beliefs work against the best interests of the believers. Vitamins illustrate how pervasive such effects can be.

Vitamin Fever

We have all heard that we need vitamin C in our diets to remain healthy. This is true. If we fall below a certain average amount of vitamin C in our daily diet, we will have health problems. But how much is enough? My drugstore has vitamin C pills ranging in doses from 500 to 1500 mg. of vitamin C. Many millions of consumers take two or more such pills daily to ward off potential ailments. But how much do you really need? Ask a few friends and they will probably say around 500 mg. In contrast, adults actually need between 75mg and 120 mg (nursing mothers), with smokers needing an additional 35mg.[18] Moreover, most people easily ingest that amount from their normal diets. Even a large helping of potatoes will do the trick. Yet, many people ingest several thousand milligrams of vitamin C pills each day, and at considerable expense.

Vitamin overconsumption reflects a widespread form of pseudoscience that also includes nutritional supplements. Here, we will follow the official practice of classifying vitamins as organic molecules that, in small amounts, are necessary for an organism's health and must be taken in through the environment because they cannot be synthesized directly by the organism. Supplements are inorganic versions, such as calcium. However, in recent years, vendors often blur this distinction by referring to the same products both as vitamins and as nutritional supplements.

Supplements and vitamins can sometimes be critical to good health. Indeed, for almost all advertised vitamins and supplements, a small, usually tiny, percentage of people do need to take them for health-related reasons.

People with lopsided or restricted diets, such as early-eighteenth-century British sailors and famine victims, need vitamin C to prevent scurvy. Folic acid and other supplements are critical to successful pregnancies for some women, and calcium supplements help other groups. But these legitimate cases constitute only a small percentage of the vitamins and nutritional supplements that are purchased and consumed every year. In 2019, US sales exceeded thirty-two billion dollars, with every indication of continuous increases in every succeeding year.[19] At least 70 percent of the US population in every age group has used some kind of supplement, and roughly 50 percent of US citizens consume vitamins and supplements on a regular basis year after year.[20] Financial analyses of the market potential suggest that dietary supplements and vitamins are one of most promising sectors for future investments by major companies and private equity groups.[21]

All of this would make great sense if vitamins and nutritional supplements were highly effective for most consumers. In reality, they rarely provide any health benefits.[22] Multiple reviews show few positive effects.[23] Even worse, the main reasons consumers take vitamins are mistaken, namely to increase general health and to provide a defense against illness. In reality, the only empirically supported reason to take a specific vitamin is to address a specific deficiency.[24] In some cases, these products are toxic in large amounts. Adverse interactions with prescription drugs present other risks, especially because patients routinely fail to disclose vitamin and supplement use to their doctors.[25]

Few studies have examined why people are so attracted to vitamins and supplements. Above and beyond centuries-old consumptions of worthless compounds to ensure health, the obsession with vitamins may have been turbocharged by the widespread publicity of Linus Pauling's claims about vitamin C.[26] Especially appealing were Pauling's claims that it could help reduce common cold infection and speed recovery. Pauling was one of the most distinguished chemistry researchers in history. He was awarded the Nobel Prize in chemistry, as well as a Nobel Peace Prize. However, he also made widely publicized claims about vitamin C's health benefits that all turned out to be unverifiable. Nonetheless, vitamin C sales took off and remain high to this day. Vitamin C allegedly protected against the common cold, atherosclerosis and angina, and cancer among many other diseases. All later carefully controlled random assignment studies showed no possible effects of high-dose vitamin C. But the beliefs in the healing power of vitamin C continue.

The biggest consumers of vitamins and food supplements are the most educated members of our society—those who have education after college. Elderly people and women are also heavier users. Many advertisements in the popular media target these groups in vitamin and supplement commercials.[27] Better educated individuals may be more likely to know about those rare cases in which vitamins and supplements are needed and then commit the "just in case" fallacy themselves. They also are more able to handle the expense. Some of the largest pharmaceutical companies have quietly bought out vitamin manufacturers. These companies know full well that these vitamins have no efficacy but welcome the easy and reliable profits that come from selling them. Vitamins and nutritional supplements have a major advantage over drugs. They are not regulated by the FDA and they do not require expensive trials to bring to the market.

Several factors support beliefs in the efficacy of vitamins. Here we focus on cognitive biases associated with children's early beliefs about successful cures. Kristi Lockhart led a project in our lab on early beliefs about how medicines work.[28] While that research did not focus on vitamins, the findings naturally extend to most vitamins and supplements. From those studies, four examples illustrate how early cognitive beliefs about cures can continue to influence older children and adults: First, ingested or injected medicines are seen as more effective for internal maladies than surface lotions or remote "cures" such as a special healing lights. Second, young children embrace an unwarranted agency for medicines. They tend to see oral medicines such as pills going mostly, or even exclusively, to the afflicted, or at-risk area of the body. If one has an infected toe, the antibiotic goes there. Pills are like heat-seeking missiles that "know" where the problem is, go there, and fix it. Third, children tend to see these agents as universal problem solvers able to fix almost any problem once they get to the afflicted site. Finally, young children think that side effects of medicines should resemble aspects of the illnesses and its typical symptoms and that side effects should occur near the site of the disease.

While these beliefs decrease with age, versions are still present in adults. Thus, medicines and vitamins are often advertised as "targeting" or "going to" problem or vulnerable areas, which resonates with early notions of agency. Marketers use the word "target" in ways that suggest the vitamin knows where it is supposed to go in the body and knows how to fix the problem when it arrives or how to construct a preventive shield against future invaders.

Many vitamins are taken for general health reasons rather than for specific deficits, an idea that is supported by early beliefs in medicines as all-purpose "problem-solving" agents. For an example of targeting's survival in adult reasoning, a recent Twitter thread included the following targeting-assuming question: "How does Advil know what part of your body hurts?" In the thread was a response that Advil "goes into the bloodstream, then to the brain, and then follows the pain signals down to nerves to where it needs to go." It is such a cognitively comfortable narrative even as it is granting Advil pills an alarming degree of agency, and even intentionality.

Finally, advertisements exploit real or contrived similarities between products and health conditions. A popular advertisement for an alleged brain supplement shows a glowing jellyfish and a brain, glowing in the same color. The compelling suggestion is that taking an extract that helps jellyfish glow will help your aging brain glow more like a youthful one. Yet, glowing brains from images of brain scans are computational

Figure 8.1

The evolution of vitamin marketing. The first marketing efforts in 1922 promised vague unproven effects, such as increasing health, vigor, and beauty. (The word "vitamine" first appeared in the scientific literature in 1912.) More dramatic claims emerged later. In 2021, several hundred vitamin "cure" books were purchased by eager readers. A few typical titles of the three most common types are shown. As of August 2021, over a dozen new books described how vitamins could help fight off COVID-19.

Source: 1922 marketing brochure cover courtesy of John P. Swann, PhD, Historian, FDA History Office.

reconstructions of blood-flow patterns. Brains do not actually glow. The advertisement exploits the silliest of surface similarities that make no sense at a deeper level. Excessive attention to surface similarities is yet another bias that is typically stronger in children but continues to have some influence in adults.

For over 100 years, vitamin marketing has misled the public (figure 8.1). We don't fully understand why most vitamin consumers are wasting their money and sometimes causing health problems even as they are especially well educated. In many cases, latent cognitive vestiges from childhood may be working behind the scene. The over thirty billion dollars wasted each year in the United States alone could find much better uses.

Some people do resist the vitamin scams. They sense that something is off, dive a little deeper into the alleged scientific evidence, and find there is nothing there. To my knowledge, no studies have focused on these people. However, given what is known about how we can learn to manage other developmental vestiges, it should help to ask questions such as the following: "Why are these vitamin pills and supplements so essential today when they weren't available for most of human evolution?" "How does vitamin C actually protect someone from disease X?" "Do those making the scientific claims have any conflicts of interest?" We don't have to be practicing scientists to ask such questions; we just need to appreciate the importance of evidence, actual mechanisms, and some vetting of alleged experts. If preschoolers can spontaneously understand the value of such questions, we should be able to maintain and further develop those abilities as adults.

Motivations, Emotions, and Social Cultural Influences

Pseudoscience is often supported by motivational influences, such as the desire to be accepted by a group, to follow a religious belief or a political doctrine, to confirm a certain image of oneself, or to keep one's internal beliefs consistent.[29] If a scientific position is ego threatening, that threat can create a high barrier to acceptance. If you are a loyal donor to a nonprofit that opposes all forms of herbicides and strongly supports organic foods, you are more inclined to believe negative claims about an herbicide such as Roundup. You can be biased in the evidence you seek and how you evaluate that evidence. Motivated reasoning influences views ranging from climate change[30] to genetically modified food.[31] It emerges any time that your core beliefs or desires are threatened. Moreover, it works outside of awareness. People can confidently believe that they are being impartial and rational when in fact they are heavily influenced by internal motivational factors.[32]

Children are cursed not only by motivated reasoning,[33] but also by especially large difficulties recognizing it in themselves and others. Children lack a healthy cynicism when evaluating statements made by those who stand to benefit from the statement.[34] If a runner in a race with a big prize claims he won in a photo finish, most eight-year-olds are less likely to believe him than if he claims he lost. Younger children don't see why they should doubt more the one whose claims align with his self-interests. Not till around age seven do children become cynics who understand how motivated reasoning biases judgments and statements. Even more dramatically, Kristi Lockhart showed (chapter 3) that young children don't see blatant boasting as an act of self-promotion but rather as a generous offering of useful information to others.[35] These developmental patterns suggest that when under cognitive load, such as shortly after diagnosis with a life-threatening disease, even sophisticated adults may revert to earlier and easier modes of processing. Those modes make motivated reasoning less apparent, and, as a result, they leave the individual even more vulnerable to pseudoscience.

Motivational effects are closely related to influences of consensus and culture. For many years, researchers have documented people's strong desires to fit in with the group. The most extreme cases were Solomon Asch's classic studies in which an unwitting participant sat at a table with several confederates who all claimed that a longer line of a pair was actually shorter.[36] My law school colleague Dan Kahan and his collaborators have shown consensus effects with climate change.[37] But the effects can occur elsewhere.[38] Suppose you grow up in a tobacco farming area or in a coal mining area. Almost everyone you know thinks the cancer risks of tobacco smoke are hoaxes by liberals or that global warming is unrelated to burning coal. Under such circumstances, you are strongly inclined to adopt the same beliefs. These group effects are rarely explicit decisions but rather are "absorbed" from one's community. This is why it can be so upsetting to some parents from "red states" to see their child go off to a liberal college. Their children join a new culture, absorb the local consensus, and then return home over the holidays to clash with many closely held beliefs in their hometown. This was my experience visiting home as a college student during the Vietnam War.

Kahan has illustrated how the social group that one identifies with (e.g., East Coast liberals, fundamentalist Christians, coal miners) can predict people's position on a scientific debate, even as they might be convinced that they arrived at their conclusions on their own. When the strongest believers are also the most science literate, they can become dangerous purveyors of distortion. Deep mechanistic understanding can enable reflective, analytical, and well-educated people to rationalize and convincingly argue for profoundly mistaken views. Moreover, they can earnestly believe what they say. Climate change appears to be an especially strong case in which deep mechanistic knowledge doesn't always provide protection from pseudoscience, and Kahan has found similar effects with evolution.[39] These two cases may share strong identifications with a political ideology and perhaps even religious doctrines.[40]

More generally, your surrounding culture can cause you to distrust and even deny the scientifically accepted mechanistic account. Instead, you may distort that account into a mechanistic version that is more consistent with your culture. But even in these cases with eloquent zealots who are technically sophisticated, there is hope. Good-quality well-grounded mechanistic explanations usually move the public more toward good science than away from it.[41] It is difficult to maintain a dogma-driven false mechanistic alternative when it is directly, and patiently, confronted by a carefully laid out true account with easy to understand links to evidence. As researchers learn more about the best ways to provide legitimate explanations to the public, these positive effects will only increase in effectiveness.

Lysenko—Pseudoscience, Politics, and Power

An especially negative effect of motivated reasoning and cultural consensus occurred with the Trofim Lysenko's rise to power in the Soviet Union. Lysenko was a Ukrainian peasant who thrived under Stalin's protection. He championed a twentieth-century version of Lamarckian theory, in which organisms adapted to their environments while growing and then passed that adaptation onto their offspring. Based on some of his own plant research, which was later considered a combination of fabricated results and flawed methodology, Lysenko argued that growing wheat plants in the right environment could transform them into almost any needed strain. They could acquire new characteristics if you were just clever enough in how you brought them up. This included the terrible idea of growing them too close together so they could better "communicate" with each other. Following Lamarck, Lysenko assumed those acquired properties were then passed on to the next generation.

Lysenko claimed his method would result in more abundant crop yields even in soils that were normally inappropriate for wheat. Stalin embraced Lysenko's ideas as they seemed compatible with Soviet ideology: put growing organisms, including people, in the right environment and you can shape them into anything you wanted and then pass those acquired characteristics onto the next generation. This approach was adopted despite thousands of studies throughout the world discrediting inheritance of acquired characteristics and supporting Darwin's idea of adaptation through natural selection.

Lysenko was appointed as director of the Institute of Genetics in the USSR's Academy of Sciences. He was soon giving large public lectures denouncing Mendelian genetics and Darwinian evolution and all biologists who allied with that point of view as well.[42] Those biologists were routinely fired from research positions. Thousands were sent to prison camps and psychiatric prisons, where many died. The most famous case was the eminent plant researcher Nikolai Vavilov, who died after a few years in prison (see figure 8.2).[43] Vavilov had assembled the world's largest collection of seeds in the Leningrad seed bank. After his imprisonment and death, during the siege of Leningrad (1941–1944), several of his colleagues guarded large bags of those seeds from all over the world. They refused to eat the seeds because of their value to future science. Nine of them died of starvation with life-saving bags of grain nearby. Fortunately, many of those seeds are preserved to this day.

Lysenko's agricultural practices were implemented on a grand scale and were a colossal failure in which almost all wheat plants perished, resulting in widespread famines where millions died. In the early 1950s, Mao's China also embraced Lysenko's practices because of the political harmony with their own doctrines. Several more millions died from the resulting famines. Lysenko, with Stalin's and Mao's support, was responsible for many millions of deaths because of his shameless advocacy of pseudoscience.[44]

Figure 8.2
Nikolai Vavilov, founder of one of the world's largest seed banks, died in prison because his beliefs clashed with the pseudoscience of Trofim Lysenko. The photo shows him in prison shortly before his death. The stamp gives a sense of how long it took for his country to recognize his contributions to biology.

Lysenko's pseudoscience has had surprisingly long-term reverberating effects. When I visited the Soviet Union in 1984 as part of a scientific delegation of developmental psychologists, the immediate rock star of our group was an expert on behavior genetics. Our hosts explained that they were desperate to catch up because they were still so far behind in research due to Lysenko. Unbelievably, in 2017, a paper trying to rehabilitate Lysenko appeared in the *European Journal of Human Genetics*. The authors suggested that the new science of epigenetics was what Lysenko really meant. Epigenetic effects are very different from Lysenko's theory and the paper was soundly rebutted by several experts,[45] but Lysenko's ideas seem destined to survive.

Others cases of cultural influences overriding science with life-threatening consequences involve denials of medical research. One egregious instance occurred when Thabo Mbeki, president of South Africa, resisted the use of the antiviral drug AZT for treatment of AIDS. Mbeki had apparently stumbled across a paper by two US researchers, Peter Duesberg and David Rasnick, who claimed, in a marginal journal, that AIDS was a result of poverty and associated malnutrition. They dismissed the human immunodeficiency virus as the cause, despite massive scientific support. Rasnick had also worked for a vitamin company that sold vitamins to treat AIDS.[46] Mbeki blocked the use of AZT and other antiretroviral drugs to AIDS victims in South Africa and supported a vitamin regime to treat the disease, a policy that resulted in hundreds of thousands of deaths.[47]

Mbeki's decisions seem to have arisen from a confluence of cultural consensus and motivated cognition. For historical reasons many people in sub-Saharan Africa had a deep distrust of large multinational pharmaceutical companies. This distrust created consensus that blocked the science. Mbeki also engaged in motivated cognition. He wanted to act in a way that was consistent with his own self-image as a thoughtful reader of original sources who was not merely swayed by others. Similar processes may explain the anti-vaccine movement.[48] Recently, in many parts of the world, some leaders have distorted the science concerning the COVID-19 virus so as to advance political goals and profits instead of protecting the public.

Because motivation and consensus/cultural effects have led to pseudoscience with lethal consequences and because such effects can enable those with science literacy to make more convincing cases for pseudoscience, might it be better to suppress teaching too much knowledge? Some argue that we shouldn't expose the public to the details of how things work because that information provides some of the smarter extremists with better tools to hijack the message toward some bizarre conspiracy theory.[49] Those tools enable them to construct more convincing falsehoods in service of political agendas that can have disastrous effects on public health and the environment.

Such negative and demeaning views of the public are misguided and can at best only reference a few rare cases in which more knowledge enabled greater distortions.[50] Limiting public understanding is like adopting a policy ensuring that people don't eat too much protein because it allows bullies to become too muscular, or that we shouldn't teach artificial intelligence (AI) because it enables despots to suppress uprisings. In most instances, well-designed, well-justified explanations move the public toward embracing the best scientific knowledge, not toward rejecting it.

When No Mechanism Is Available

When we have no notion of mechanism at all, are we helpless in evaluating scientific claims? In areas such as quantum computing there may be no easily accessible models of how things work. Even without such models, we can use other tools from childhood to evaluate claims. One strategy gauges the diversity of sources. If one group of sources making a claim are all from the same background and organization while another group of sources making a contrasting claim come from different backgrounds and organizations, all else being equal, the claim made by the diverse group is likely to be more accurate. The caveat "all else being equal," however, matters. You need to be confident that the overall quality of both groups of sources is roughly the same. For example, if you encounter a claim made by three reporters from the *Wall Street Journal*, and another claim made by reporters from the *Wall Street Journal*, the *New York Times*, and *Al Jazeera*, you would likely be safer with the second group because they agree despite coming from very different backgrounds. If, however, the second group was a diverse set of gossip tabloids, you wouldn't be swayed by diversity.

Sometimes, such as with Lysenko, alleged high-status authorities can force false science on everyone else. People also tend to forget negative indicators of source quality and just remember what the sources said.[51] To protect against this, it can help to rephrase the source information in ways that don't repeat the statement. For example, instead of "It is myth that high-dose vitamin C helps prevent the common cold," you might say, "All available evidence shows that high doses of vitamin C are useless and can be harmful." General science literacy and numeracy increase detection of good sources from a larger set of uneven quality.[52] Even when you don't have any mechanistic knowledge and have no access to other sources, you can ask sources to mechanistically explain their point of view. Dogmatic self-appointed experts typically have difficulty unpacking their position into a coherent mechanism; they either end up with internal contradictions and inconsistencies or default to vague and unsupportable assertions.

What I Should Have Said and Asked on That Airplane Flight

Returning to my seatmate on that unsettling cross-country flight, I could have done better even with my ignorance. I might have asked him what it would take to change his mind about climate change. What pattern of data, what feasible experiment, what sources would he find convincing and why? If he claimed nothing could change his mind, I should have immediately had much stronger reasons to doubt him. If he did list details and/or feasible studies that would matter, perhaps we could have together designed the perfect study. In doing so, I would have also gained a better sense of the extent to which he was willing to change his own beliefs and engage in genuine wonder or whether true wondering and willingness to have his views transformed had left him long ago.

I should have brought up analogous cases that seemed to support small amounts making big differences. For example, I might have asked how many parts per million of pollution or pollen could have massive effects on visibility and health. I didn't know those numbers, but I had a sense that they were small, and I am guessing he knew that as well. I would then have asked why CO_2 was so different. Probing carefully contrasted cases can be very effective. I might also have politely told him about motivated cognition and how it could work outside of awareness and then asked him if he thought he was immune from such influences. I am almost hoping to have another equally formidable seatmate in the future on whom to practice my skills at getting to the truth with minimal domain knowledge. My skills certainly need a lot of work.

How Wonder Matters

If parents, children's book writers, teachers, science museum exhibitors, and other adults view young children as inhabiting a much cruder concrete stage of conceptual development, that attitude can lead to ineffective science instruction and suppression of wonder. If those adults also have negative attitudes toward science, those attitudes may easily infect the children around them. Finally, if left unaddressed and without alternative conceptually deeper models, the cognitive strategies and biases that enable infants and young children to gain cognitive footholds in understanding the world can become crippling cognitive legacies. As such, they work against acquisition of more mature and sophisticated scientific theories and beliefs. When all of these negative elements come together in the same person, they not only stifle early wonder but they foster pseudoscience, disengagement, and even outright hostility toward science.

Gaps in understanding pose additional problems. Virtually everyone's explanatory abilities run out of steam in most areas. As my son Marty remarked about his medical school studies, the deeper you probe about how diseases cause symptoms, the more you either defer to others or realize that those levels of mechanism remain unknown. When we defer to other individuals and communities, we need to understand how those people and groups are working with partial understandings themselves. When we don't, we are more easily misled. Climate change has many lingering explanatory holes, and many of us down the information chain are unaware of these gaps in full understanding.

Distrust, disengagement, and denial can create a wonderless life. They reinforce each other and lead to a decline in learning and a vulnerability to misleading biases. These three Ds emerge in childhood and can grab hold of entire communities. They can distort science toward only the crudest and most immediate benefits. While few question cases in which a discovery improves health or quality of life, many are biased toward only the most obvious and short-term outcomes. This impatience contradicts countless examples of how basic science and long-term technology projects have enriched future generations. No one could have anticipated how an eccentric monk's meddling with peas in a remote monastery garden could be a critical step toward the development of twenty-first-century genomics, or how Ørsted and Faraday's discoveries of the links between magnets and electrical current in the 1800s could lead to the ability to visualize thoughts and feelings in real time in the brain through functional magnetic resonance imaging scanners.

Distrust in science is everywhere. In biomedical research, even the most eminent figures at the most hallowed institutions, including my own, are castigated for conflicts of interest. It doesn't help when some of those individuals accept drug company funds (often through discrete and circuitous routes meant to hide the transaction) for making favorable remarks or selectively publishing positive findings about a company's products. Everyone knows about the drug that was pushed relentlessly in the media that turned out to be harmful with apparent foreknowledge by its maker. But these cases are much less common than many believe. The vast majority of scientists are careful, thoughtful, and ethical. Nonscientists from all walks of life should try to get to know scientists at a personal level and spend time with them in venues where their lives intersect. When a scientist works at a university or research center in a small town, this happens organically through sports teams, volunteer groups, childcare centers, and other settings where everyone is involved. In more populated places, we may need to make more deliberate efforts to get to know scientists.

In many cases, distortions are enhanced by suggestive but false equivalences of groups that disagree. Evolution is often cast in a "balanced" debate between traditional biologists and creationists when in fact the evidence for evolution is overwhelming and the evidence for intelligent design is nil. Climate change is often seen as equally countered by scientists who disagree when, in fact, few cases in the history of science have achieved greater consensus on one side of the issue. People are willing to doubt and even condemn those who have spent their entire adult lives in laboratories and deep study. Distrust in science seems to wax and wane in large sectors of the population as motivational factors bias judgment.[53] These false equivalences between large and tiny groups of researchers that disagree are harmful. They reinforce illusory consensuses.

Distrust is often accompanied by disengagement. As people come to distrust what they encounter in the media and through their various curated news feeds, they disengage from serious attempts to actively learn about and explore their world. To engage means far more than appreciating the "magic" of a new gadget, such as those globes that "float" on desks everywhere. Engagement is deliberate sustained inquiry. It requires a desire to find out how things like gravity-defying globes work. Engagement is supported by environments that welcome the why and how questions that come so naturally to all of us when we are unencumbered by fear of being censured or shunned for asking them.

Disengagement welcomes denial. It takes the form of selective denial of any scientific discovery that doesn't fit with one's own desires and worldviews. We all too often witness selective refusals to even entertain the possibility that some scientific claims could be true. Distrust and disengagement make denialism possible because one is unable to even evaluate most claims in any reasonable manner and instead must simply deny, often relying on the inferred and sometimes illusory consensus of one's own group.

The three bad Ds have resulted in distancing the public from science, but, surprisingly, they have not yet led to an overall desire to stop federal funding for science. For largely practical reasons, most people in the United States still favor increased funding for science to keep the United

States ahead of competitors in military technology, to keep citizens healthy, and to slow the declines associated with aging.[54] These anxieties, related to defense, disease, and deterioration (three more bad Ds), skew science approval toward immediate self-serving gains with little appreciation of the basic research infrastructure needed to support those gains.

Is the need for reawakening wonder slowly declining anyway? Even though journal articles may continue to pile up at ever-increasing rate, these patterns of increasing productivity are driven by forces completely distinct from public interest.[55] On all inhabited continents, reports proliferate in which students and the public at large are described as having declining serious interests in science.[56] (Antarctica is an interesting exception as almost all its denizens *are* scientists.) In all other areas of the world, beyond the labs and research centers (and sometimes even in them), lack of sustained interest encourages gross misunderstandings of science. Many people from diverse cultures falsely believe that safe vaccines are harmful, that infectious diseases such as AIDS and COVID-19 are hoaxes, and that astrology is a legitimate realm of physical influences on the body and the mind. The president of the United States in 2019 claimed that windmills can cause cancer in those who live nearby and did so with a certainty that preempted further discussion.[57] In 2020, he proposed injection of disinfectants into the body to counter the COVID-19 virus.[58] We need to address the factors that lie behind these widely distributed failures to acknowledge the results of scientific research and to ignore the most basic scientific principles. Proliferation of publications, and funding of sciences, do not automatically guarantee a better informed public.

People do know some surface trappings of science. A 2019 Pew survey approvingly remarked that a majority of US citizens did quite well on its science test.[59] But what was in that test and what was missing? The test items covered such things as knowing that oil and coal are fossil fuels, detecting an appropriate control condition in a very simple experiment, showing how to read a bar graph, and identifying the definition of an incubation period. There were no questions about how things actually worked that revealed a deeper sense of mechanism. But senses of mechanisms and explanatory causal frameworks and fragments may be the most critical ingredients in maintaining an ever-growing sense of wonder. Well-developed wonder in turn drives the greatest appreciations of basic science and the ability to evaluate new discoveries and claims.

For many of these negative scenarios just described, early engagement and pleasure in learning science offer powerful inoculations against bad science and lack of interest in science as a whole. We protect ourselves from malicious myths and pernicious propaganda by keeping wonder and its associated cognitive skills alive and flourishing. We also gain an inspiring rather than a deflating form of intellectual humility, an eagerness to argue to learn, and an appreciation of having our minds changed by better ideas. How do we move in that direction?

Putting aside motivations arising from defense, disease, and deterioration, public interest in more basic science depends on the adoption of three good Ds: doubt, deliberation, and deference. These helpful practices can only arise from a better understanding and engagement with science. First, it is appropriate to have a healthy degree of doubt about virtually any new scientific claim. Doubt in this sense does not carry the negative tone of distrust but rather is a recognition that there is room for both human and machine error in even the most carefully and ethically conducted studies. Some dramatic "discoveries" have turned out to have fatal flaws. Famous cases include cold fusion, bacterial life on Mars, and alleged links between electromagnetic fields generated by power lines and cancer.[60] Doubt can also have a gentle influence on theories by working with wonder to refine and continuously improve a roughly correct account with some details that don't seem quite right. Many laboratory group discussions work in precisely this way and with an atmosphere where doubt is both encouraged and welcomed.

At least a small element of doubt should be present when encountering new discoveries, especially when they challenge decades of prior work. But instead of leading to denial, doubt should lead to deliberation about the feasibility of the new claim. A sophisticated "doubter" can often get quite far in evaluating new claims and, in doing so, may often rely on skillful deference. Well-developed doubt involves careful consideration of the claim, how it is justified, and how much it disrupts other known "facts." With respect to deference, our trust in a claim may depend on who made the discovery in terms of both their level and area of expertise. In many cases, after careful deliberation, a doubter might realize that he or she does not fully understand the claim and/or the study but is nonetheless quite confident in his or her decision to defer to certain others about the claim.

Despite the apparent complexity of these skills, young children show sophisticated patterns of deference. We all should nurture a full suite of good D skills that enable a lifelong appreciation and enjoyment of science. In doing so, we benefit ourselves and our societies as a whole. We should practice these skills early and often and use them to build a body of knowledge and understanding that can override biases, even as we might still fall prey to them when rushed or distracted.

An additional benefit to being engaged with a broad range of scientific communities is that engagement enables us to gain a better sense of those communities themselves. This helps us navigate challenging topics such as climate change. After the encounter with my eloquent and sophisticated airline passenger, I was motivated to dig deeper because of the many other people I knew who were involved in the relevant sciences. I knew they were not all liberals, bicoastal elites, and that some had even changed their minds. This awareness made me less likely to think they were driven by a single political agenda. All of us can benefit from a greater understanding of how scientists are trained and how their best work is evaluated. We can use that information to better interpret what is behind a group's claims.

Wonder does not provide perfect protection from faulty theories and misplaced mechanisms. Outstanding scientists have made enormous blunders even as they have made remarkable discoveries. Consider the case of the brilliant thermodynamics scientist William Thomson (also known as Lord Kelvin), who once said that "I am never content until I have constructed a mechanical model of the subject I am studying. If I succeed in making one, I understand; otherwise I do not."[61] Despite a string of fundamental discoveries in physics, Thomson was also convinced that airplanes could never be practical and initially dismissed the discovery of X-rays. To his credit, when he read Rontgen's paper on X-rays, he wrote to Rontgen, ". . . when I read the paper I was very much astonished and delighted. I can say no more now than to congratulate you warmly on the great discovery you have made."[62] But Thomson did not change other mistaken beliefs such as his rejection of evolution. Other great scientists have clung to their views despite a steady stream of counterevidence, such as Linus Pauling with respect to his beliefs about the health merits of vitamin C. Why didn't the wonder that propelled them to great discoveries also protect them from clinging to bad science?

A major reason may be the difference between truly "range free" wondering, which deliberately tests boundaries, and "fenced in" wondering, which constrains wonder within the confines dictated by dogma, motivated cognition, consensus, and intimidation. No one is immune from thwarting their own remarkable powers of wonder if they are not willing to doubt their own beliefs and to repeatedly adopt a reflective view of what they believe and why. The most potent forms of wonder co-occur with a strong intellectual humility in which we are always willing to acknowledge our own potential fallibility. Having an open mind does not mean a weak "pushover" mind, just one that is always willing to wonder about its own limitations. Even the best of us have to continuously keep this in mind or risk sliding into a rut of inflexibility and denialism.

▪ ▪ ▪

A life without wonder happens when its early blossoms are crushed, leaving a void in which mere consensus, motivated reasoning, and cognitive biases take over. People can come to adopt views and embrace accounts for all the wrong reasons. A wonder-filled life offers a more positive future. Enriched public understanding of science and technology, coupled with attendant practices of appropriate deference, have potentially enormous social benefits. We can all maintain or reawaken the thrill of discovery and insight. We can see more clearly how advances in all areas of science and engineering enhance and enrich our lives. We can also better anticipate what directions of research should be undertaken with great care and caution. To get there, we need to consider more concrete courses of action.

9 Great Reawakenings

I once heard a colleague from Jamaica wax euphorically about a visitor to their medical school in Kingston some thirty years earlier. My colleague was praising Alexander Thomas and Thomas's insatiable and sophisticated curiosity about everything and anything he encountered during his visit to Jamaica. I had never met Thomas, but I was well aware of his research with Stella Chess on childhood temperament and personality.[1] I knew nothing, however, about his own personality and about his apparently joyful, yet relentless, lines of questioning. He wanted to know why certain crops, such as coffee, thrived only in particular locations, how the historical Spanish presence had influenced the Jamaican language, why none of the Jamaican snakes were venomous, and why Jamaica had crocodiles but no alligators. He had countless questions not only about cultural and legal topics unique to Jamaica but also about why certain human diseases were prevalent. He was described as a delightful mix of an exuberant child and a brilliant scientist, bubbling over with curiosity and questions, but also wonderfully sophisticated in his chains of questions as he dove deeper and deeper to gain better understanding. I remember this conversation so clearly because I felt such admiration and awe for someone whose "discovery mode" was always "on." I like to think of myself as inquisitive but knew all too well that I can easily slip into more passive modes of mindless entertainment. I often think about how many pleasures of insight I may have missed because of these "off periods."

We all have encountered people like Thomas. The polymaths discussed earlier were almost invariably described in a similar manner. We know from these people that lives of nearly constant inquiry and insight are possible, but we also know how often it doesn't happen. What might be done so that all of us spend more of our time exuberantly developing deeper understandings of natural phenomena and the engineered world? In describing some possible courses of action, we should also again address why we might all want the rewards of spending more of our times in such states.

What benefits arise from reawakening wonder, and how do we get there? Prior research directly supports the feasibility of most steps proposed here. When it doesn't, those steps conform with current views of cognition and its development. Starting at the family level and then spreading to larger and larger groups, our focus is not on science education in the schools but rather on what happens everywhere else. As one large group of researchers declared, "School is not where most Americans learn most of their science."[2] It's also not where they learn to love science. Children engage in exploration of mechanisms well before schooling, and those tendencies need to be appreciated and nurtured. Yet, adult naive beliefs about children's minds often clash with how young minds actually work. Such interpretive conflicts pervade entire societies' views of children and how they learn. Accordingly, we need to think about changes at all levels from individuals to large-scale communities.

The societal benefits of science literacy have been a constant topic of interest. A recent National Science Board survey of science attitudes in the United States concluded rather morosely that the United States had fallen from the sole leadership position to being one of many leading nations.[3] My sense is that most people don't care that the rest of the world is getting better; rather they worry that the United States is losing its love of science and technology. Sometimes we seem to be the Sputnik era in reverse.

Historically, advances in science had enormous mostly positive effects. In his book *Enlightenment Now*, Steven Pinker describes how the Enlightenment bettered societies through its elevation of science and reason.[4] Advances can improve health; make work easier, safer, and more pleasant; and enhance leisure and recreational activities. Consider just six inventions and discoveries from the twentieth century: insulin, antibiotics, the lowly shipping container, transistors, LED lighting, and lithium batteries. Innovations in these areas was often based on work in basic sciences that then had far broader consequences than ever imagined. Insulin has saved and improved the lives hundreds of millions of people. Many more millions of lives have been saved through use of antibiotics, even as their heavy use may now be breeding "superbugs" that may soon be resistant to all known antibiotics. "Containerization" was a multidecade process of small incremental inventions (e.g., locking mechanisms between containers) that ultimately led to 90 percent of all cargo worldwide (other than minerals and chemicals and other bulk cargo) being transported in standard shipping containers at greatly reduced costs.[5] Transistors made possible circuits with billions of times greater complexity than prior tube circuits. Without transistors, there would be no cell phones, personal computers, or modern cars. Lighting approaches around 7 percent of the total energy consumption in developed countries. If all lighting were switched to the most efficient LED forms, those costs might go below 1 percent of total energy costs.[6] Lithium ion batteries are considerably lighter than traditional lead acid batteries while also charging more quickly and having much longer lifespans. Without them, cell phones, laptops, and electric cars would not be feasible.[7]

Many other scientific discoveries and inventions have changed millions of lives but are far less familiar. My favorite little-known discovery is the Haber-Bosch process. In 1910, Carl Bosch, following a 1909 feasibility demonstration by Fritz Haber, created the first large-scale method for producing ammonia. (Haber and Bosch both received Nobel Prizes in chemistry.) The process converts nitrogen (N_2) to ammonia (NH_3) by a reaction with hydrogen in a complex procedure involving a catalyst, high temperatures, and high pressures.

The agricultural impact of synthesized ammonia was enormous. Ammonia is a key ingredient in fertilizer. Fertilizers can increase crop yields by as much as four times the harvest per acre. Given that in many parts of the world well more than 50 percent of the arable land is under cultivation, without the Haber-Bosch process, those areas would likely experience mass famines. By one estimate, roughly half the world's current population (almost four billion people) exists today because of the Haber-Bosch process.[8] However, large amounts of fertilizer run off from crops into the water system and surrounding soils, causing unanticipated effects such as algal blooms. The Haber-Bosch process also results in increased CO_2 emissions because of the large amounts of energy it requires.[9]

Ammonia was also central to the manufacture of military explosives especially since natural supplies of useful nitrogen were in controlled access areas such as the guano-rich islands off Peru. The Haber Bosch process hastened Germany's entry into World War I, greatly prolonged the duration of the war, and ultimately was connected to 100–150 million deaths in various conflicts throughout the twentieth century.[10]

Science and engineering discoveries, like any tool, can have a dark side as well, creating possibilities for new weapons, methods of surveillance, and ways to control and manipulate people. Some of these developments may be unfortunate necessities in times of conflict between nations, but for many others, the best way to thwart such outcomes is to create an engaged, science-literate public.

Less Tangible but Highly Rewarding Benefits—New "Lenses"

Increased understandings of science and technology enrich our lives in more intangible ways. Just as new glasses can enhance vision and thereby increase aesthetic appreciation and the ability to navigate the world, greater understanding can enhance cognitive insight, aesthetic sensibilities, and mental mobility. A better sense of how the world works and how to buttress our understanding by accessing knowledge sources makes the targets of thought more brilliant and beautiful. As I write this on an early spring day and hear hundreds of different birds sing around me, I remember some science about birdsong. For many male songbirds, parts of their brains can double in size for just a few months to enable them to sing more gloriously and attract mates more effectively. These brain regions quickly shrink back to their smaller size after mating season because of the energy requirements imposed. I remember how their songs are "designed" to be to be maximally distinctive and acoustically robust and why their complex structures can only be acquired through intricate developmental pathways, and I recall that all of this is to attract and keep a better mate. For me, hearing a birdsong is also visualizing all the machinery that sustains it and why. I don't actually know all that machinery in detail, but I acutely feel its presence and know that it awaits further inquiry.

When a car glides by our house, because of what I learned long ago as an undergraduate, I marvel at the extraordinary triumphs of engineering and science that managed to harness hundreds of chemical explosions a minute into a well-functioning, nearly silent, mechanical system that provides smooth accelerations through thousands of interactions among constituent parts. I appreciate just how hard it was to develop a car chassis that didn't constantly vibrate apart. I also appreciate much more clearly, because of my modest grasp of climate change, that this marvelous device must soon become a museum piece replaced by greener forms of transportation.

Finally, consider thunderstorms. Large thunderstorms may have updrafts moving at 100 miles per hour and lifting over a million tons of water vapor several miles into the sky. In strong storms, this water vapor updraft can burst through the atmosphere layer known as the *tropopause,* which is where many clouds normally form their tops. These geysers in the sky surge far up into the stratosphere and can change atmospheric conditions for thousands of miles. Every thunderstorm has within it these "elevators" lifting vast amounts of weight at breakneck speeds.[11] This elevated water vapor becomes potential energy that can be released through precipitation and lightning. The energy released by a typical thunderstorm is greater than that released by the atomic bomb in Hiroshima. On average, roughly 2,000 thunderstorms occur each day somewhere on earth, the energy equivalent of 2,000 nuclear bombs exploding.

Lighting can mimic the Haber-Bosch process by bonding nitrogen to oxygen. Nitrogen dioxide is carried down in rain as nitric acid and enters the soil, acting as fertilizer. Rain without thunder and lightning doesn't have that virtue. How amazing that rain from a thunderstorm is different, and beneficial, in this way! So much is happening behind the scenes when a thunderstorm sweeps overhead.

Science and technology affect everyone's lives, including those who decry such influences in social media. But science should also enrich everyone's experiences of the world by enabling deeper, more meaningful perceptions and interpretations of reality. Those who claim to detest science are denying their own developmental roots. It makes sense to dislike some of the trappings of science or ways in which science content can be communicated in education and the media. But to dislike the insights afforded by science itself seems akin to having a dislike of a visual perception or a distaste for a good memory.

The strongest professional interest in science from the humanities usually comes from philosophers of science and historians of science. Professors of comparative literature or religion rarely try to understand a great scientific discovery or a new technology to further their work (other than the digital humanities). Conversely, while many scientists and engineers paint, sculpt, compose music, and write poetry, very few read the scholarly writings of colleagues in departments of English, art or music. Even fewer find inspiration in those writings that informs their work. We should all try harder to integrate and connect across all the disciplines so as to understand them better and ultimately inform our own work. Wonder knows no disciplinary bounds. Exceptions to these divisions are inspirational. Michael Frayn's *Copenhagen*, a play about a meeting between Niels Bohr and Werner Heisenberg, has inspired scholarly contributions from scholars in science and engineering departments as well as from sociology, political science, philosophy, history, English, and of course drama and theater studies.

Unfortunately, the main point of intersection between the sciences and the humanities swirls around morality, as have many of the papers about *Copenhagen*. This is surely too narrow a connection. Humanism may be about the moral (or, as Pinker claims, the sentiment of sympathy), but *humanities* is defined as the study of human products or processes: art, architecture, literature, music, and surely science and technology. Yet for the sciences, unlike many other products, humanities scholars often adopt a humanistic stance and focus on moral issues. But surely some of the strongest connections across these areas are through ideas. Ideas, models, and interpretations in one domain can give flight to new ways of thinking in the other. Richard Holmes, in describing the euphoric interactions between scientists and Romantic poets in his book *The Age of Wonder*, shows how the evolving concept of infinity was fueled by excited discussions between poets and astronomers in the late 1700s. Science and the humanities should interact at deep intellectual levels, not just through moral poses.

Much more recently, the writer Robert Macfarlane went thousands of feet beneath the earth's surface to talk with a physicist who was looking for the presence of tiny subatomic particles called neutrinos. After learning that enormous numbers of neutrinos passed through the earth every second, he asked the physicist:

> Does it change the way the world feels? . . . Knowing that 100 trillion neutrinos pass through your body every second, that countless such particles perforate our brains and hearts? Does it change the way you feel about matter—about what matters? Are you surprised we don't fall through each surface of our world at every step, push through it with every touch?[12]

The physicist said it dramatically changed his everyday experiences of the world, making it even more astonishing and beautiful. Macfarlane later confessed to also having a changed sense of nature and a persistent awareness that, for neutrinos, "this world is as mere mist and silk."[13] Just one wonder-inspired visit to a physicist's laboratory and Macfarlane could envision himself interwoven with the physical world in a starkly different way. We should all experience such striking new perspectives many times throughout our lives.

For the sciences and the humanities to connect in more insightful ways, all parties must seriously engage with ideas in areas far beyond their areas of expertise. Literature can still inspire scientific models. In 1973 the evolutionary biologist Leigh Van Valen introduced the "Red Queen hypothesis" by referring to the following passage from Lewis Carroll's *Through the Looking Glass:*

> "Well, in our country," said Alice, still panting a little, "you'd generally get to somewhere else—if you run very fast for a long time, as we've been doing."
>
> "A slow sort of country!" said the Queen. "Now, here, you see, it takes all the running you can do, to keep in the same place. If you want to get somewhere else, you must run at least twice as fast as that!"

Van Valen used the idea of having to run faster and faster just to keep up as a way of formulating how two species may coevolve in an arms race where each must evolve at faster and faster rates just to keep up with each other in ability to survive and reproduce.[14] To this day, the Red Queen hypothesis remains a hot topic in biology, and indeed some contemporary biologists have used other passages from *Through the Looking Glass* to further explore the hypothesis.[15] We can imagine how other current problems in the sciences, such as finding the part of a virus that cannot easily mutate as it tries to disguise itself to fool the immune system, might benefit from deeper knowledge of fiction and history. There are hundreds of fictional and historical cases of individuals who adopt elaborate disguises only to be unmasked by a telltale feature, such as a limp, that they cannot change. Knowledge of many such cases might well amplify a biologist's conjectures about which aspects of a virus are less able to mutate; the stalk of the COVID-19 virus is one current proposal. Of course, the insights run in both directions. A novelist might gain new inspirations in crafting an imposter story from research on how viruses mutate to avoid detection.

I've focused on how a deeper understanding of science and technology can increase aesthetic appreciation and not undermine it. Science adds delight, not debasement, to our appreciation of nature. Appreciating beauty in nature is not simply being awestruck and stupefied. It also involves seeing and enjoying a rich and textured causal terrain unfolding far off to the horizon. How can we ensure that the next generation has greater access to these experiences? Starting at the level of interactions within the family and gradually moving up to larger and larger scale communities, we can implement many specific and evidence-based interventions that will help reawaken and maintain wonder in everyone.

◻ ◻ ◻

Concrete Courses of Action

At the Family Level—Conversations

What you say to children, how you say it, and how frequently you engage in such activities all matter, often more than is appreciated. Conversations with children influence their own approaches to learning more about the world, both in terms of what they think are important questions to ask of the world directly and also what to think of others who know more than they do.[16] Too much attention has been paid to the alleged thirty-million-word reduction in words heard by underserved children in the preschool years. Later analyses suggest the gap is at most four million words.[17] Whatever the amount, content matters more than the sheer number of words delivered. You can't just play *The Music Man* song "Ya Got Trouble" again and again, hoping to saturate your child with the most words per minute and expecting great cognitive growth as a result.

We need to listen carefully to children's how and why questions and use our own understanding, or that of others, to expand theirs. Adults can have informative (for both sides) and rewarding conversations with children in settings ranging from the home to a science museum.[18] Information should be carefully sequenced and at the right levels of explanation so as not to overload children. Parents and caregivers who acquire these conversational skills enhance science learning and curiosity not only in children but also in themselves.[19] Ask a young child why something is the case (e.g., "Why are insects always small?") and you may become fascinated yourself. All *wh-* questions (who, what, where, why, and "how") are also "open-ended." Open-ended questions require responders to elaborate with details. These differ from questions where a simple yes or no response, or other simple choice, is easily available: "Did you have a good time at the zoo?" "Was it before lunch or after lunch that you left?" and so forth. How and why questions are best because they pull more directly for causal explanations.[20] Ask a child why deer have eyes more on the sides of their heads and wolves more on the front and you may quickly open up an entire domain of predator/prey adaptations. More generally, expert teachers intuitively know better than novices how to ask children more open-ended questions and questions that foster more exploration and learning about an object[21] (see figure 9.1).

Interactive skills of this sort do require deliberate practice and developing a deeper understanding of children's minds. We also need to treat children as active learners, not as passive vessels in which to deposit knowledge. When parents and friends do talk about science in an engaged and child-friendly manner, such conversations can be stronger predictors of later life interests in science than any other factors, including having science toys and materials, attending science camps, and joining science clubs.[22]

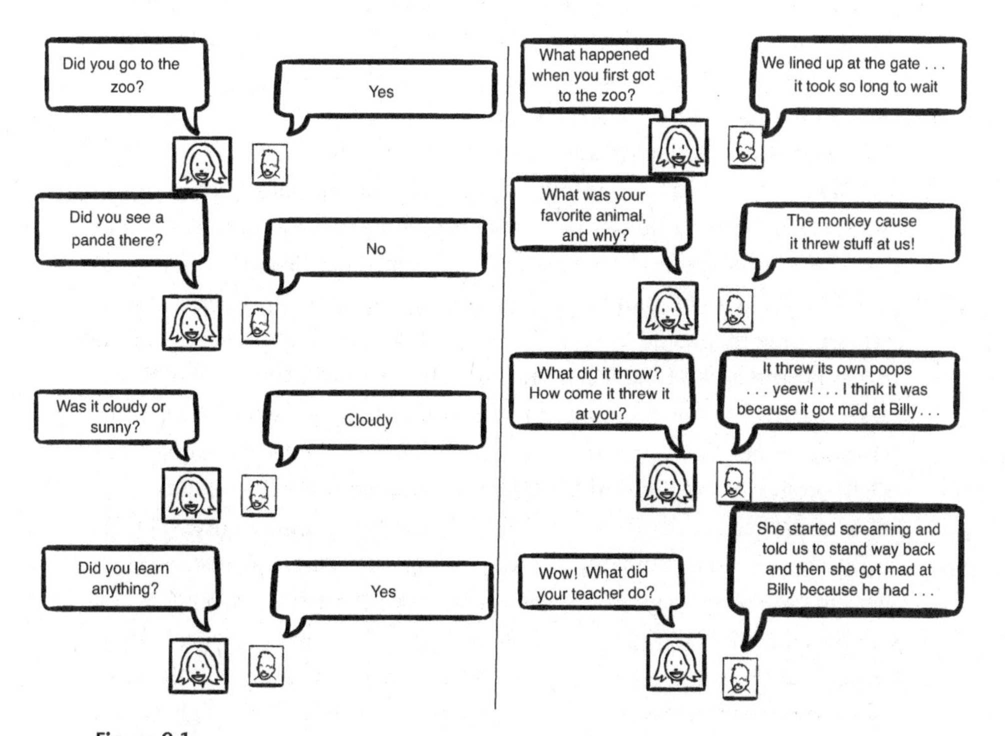

Figure 9.1
Open-ended questions. Questions that invite elaboration and more questions can transform a parent-child conversation from a one-liner (left panel) into a rewarding joint learning experience (right panel).

Certain interaction patterns can have surprisingly long-term effects. Suppose some first graders went to an aquarium and saw a demonstration of kinds of shells. Afterward, a teacher might ask those students to think more deeply about the experience, perhaps with pictures of shells as props. For example, the teacher might ask, "Which would be a better way to sort these shells into two piles—big ones and little ones or those that that have two parts (*bivalves*) or those that have one part (*gastropods*)? Why do you think the second way is better? If you weren't sure which way was better, how might you figure that out?" (The teacher could then illustrate that the more meaningful way of sorting also allowed one to see additional features shared among members of the category.)

In one study, first-grade teachers were briefly guided on how to have conversations with children that resulted in more meaningful, deeper levels of exploration and thinking. They also helped children contemplate what kinds of strategies they were using to expand their understanding. Three years later, children who had such teachers in the first grade showed advantages on memory tasks and in strategy use in the fourth grade, despite having different intervening teachers and classrooms.[23]

Nuances in adult conversations with a child can foster different kinds of responses and actions by the child.[24] Some comments promote object explorations such as seeing how gears fit together, while others promote explanations, such as describing the workings of gear mechanisms. In other cases, providing the right how and why questions can increase subsequent exploration. Imagine that an adult says, "Hmm . . . I wonder why that car is much higher off the ground than that one and why its front and back wheels are closer together," while pointing to an SUV and a passenger sedan. If the adult has gauged the child's present level of knowledge correctly, those questions might prompt the child to walk carefully around both vehicles and discover other correlated properties such as tire and suspension differences. Similar approaches in other domains, such as asking why desert plants are more likely to have waxy coatings and spikes and thorns, can lead to further explorations. Questions are especially effective when they invite thought about abstract causal relations rather than about a single causal link.[25] Instead of asking what made a bolt of lightning appear, one could ask how lightning is related to certain kinds of clouds and temperatures and why thunder sometimes is heard seconds after the flash.

Elaboration

A particularly powerful form of interaction involves *elaborations* on children's descriptions or utterances.[26] Imagine, for example, the following conversation between a parent (P) and a child (C):

P: What did we just see making noises at the beach?

C: A bird!

P: Yes . . . it was a white bird with a yellow beak and it made a sound like this [. . . parent attempts to imitate sound]

C: yeah . . . but it really sounded like this [child attempts to make screech] and it tried to eat my lunch . . .

P: Yes . . . it tried to eat your peanut butter sandwich . . . and you chased it off!

C: I did

P: So, we saw this yellow and white bird right near us at the beach on this windy and sunny day and it came up real close to you. It tried to take a bite of your peanut butter sandwich that was sitting on our favorite picnic blanket . . . and then you chased it off; and it made a funny sound like this [repeats sound].

In this conversation, the parent keeps elaborating on the child's recall of an experience and provides additional information after an initial prompt. It's a back-and-forth interaction where the parent listens carefully to what the child said and then adds additional details, some of which trigger new details in the child. They build together a richer account of the experience. Many studies find a common effect: telling parents how to elaborate in this manner improves children's memories of an event and increases children's spontaneous elaborations.[27]

Elaborations promote wonder. A parent might elaborate in an especially explanatory and causal manner. She might observe that the bird went for the peanut butter sandwich and not for the book next to it and wonder why. If the child said it was because it was food, the parent might then ask how the bird knew it was food and open up a discussion of what makes something food and how a bird could detect it and why it would help the bird to do so quickly and accurately. A parent can develop many different causal elaborations and questions that align with the child's revealed interests. Parents soon wonder themselves and become active learners, guessing and asking others for help. When adults wonder with children, questions are not inquisitions of the child but of nature itself.

Elaborated interactions do not usually happen effortlessly or with casual indifference. They normally require one person, and sometimes both, to carefully build out a shared topic of mutual interest. You must be fully engaged and not settle for simple "vegging out" or occasionally glancing up while texting. Such interactions can even be challenging between two adults. At social gatherings, we may find our conversational partners put off by deeper dives into real understanding. At most cocktail parties, things may go well enough if you join in brief conversations on topics such as new restaurants, hobbies, or alarming headlines. But if you try to drill down into the fundamentals that explain the restaurant business, the physics of sailing, or the molecular biology underlying an emerging pandemic, many conversational partners will soon back off uncomfortably. Perhaps more of such conversations should happen at social events, but it's clear how they can disrupt the other goals of such gatherings.

When parents want to nurture their children's early wonder, they can simply manifest that wonder themselves and share gratifying quests for explanations in their daily lives. When young children see their parents spontaneously engaging in and enjoying such behaviors (and inviting their children to join in), children know that they share an important value with their parents. Simply modeling curiosity about the natural world has a positive effect on a child's later engagement with science in school,[28] and modeling how and why questioning can have an especially strong effect. But getting into such frames of mind continuously is atypical today. Even in those glorious ages and communities of wonder, only a few people were in deep wonder all the time. To move toward more of that way of encountering the world may take considerable time and effort, at least at first. Only a few may want full engagement all the time, but many of us may rarely find the time to reflectively wonder at all. By neglecting this foundational part of ourselves, we can lose much both as individuals and as social companions.

Didactic Predators

Some especially earnest readers may relate to situations where highly motivated parents become "didactic predators" and take the idea of filling their child's head with science too far. They adopt the kinds of extreme didactic instruction that are insensitive to their children's cognitive states. I confess to acting this way at times only to be quickly reined in by my more socially aware spouse.

Many years ago, when we were in a park with our children, we saw two well-known professors with their three-year-old son. He had crouched to look more closely at a large grasshopper. His parents immediately leaped into action. With the best of intentions, they launched into a college-level lecturing and Socratic questioning style with their preschooler. I remember the conversation as follows (F = father, M = mother, C = child):

F: Oh look . . . you are investigating a grasshopper. A grasshopper is part of a group of animals known as insects. All insects have six legs unlike spiders that have eight legs. Can you count the legs?

C: [looks at father and gestures hopelessly at the grasshopper as counting the moving legs seems impossible]

M: Well maybe it is hard to count the legs . . . but why do you think the number of legs is important?

C: cause . . . cause . . . (gives up and shrugs)

M: It's because leg number is an especially important way to tell what kind of bug it was and where it came from in terms of its parents and their parents and so on all the way back in time.

C: Its parents? The grasshopper has parents?

F: Of course it does . . . at least in the sense of those older animals that created it. You see all animals and plants come from other animals and plants through processes called pro-cre-a-tion [sounds out loudly] and reproduction—which means making more copies of yourself. They all have inside them information written in something called DNA that carries the instructions for how to make the next animal of the same kind. And where do you think that DNA comes from?

C: Its parents?

M: Yes of course the parents . . . but how?

C: uh—[looks at grasshopper]

F: You need to know more about what DNA is and how it replicates. Let's start with an analogy to a recipe. Remember when . . . [continues expanding on recipes as codes until he notices that child has hurried off, ignoring the continuing lesson].

M: Let's follow him and see what else he gets interested in . . . we can continue there.

While my memory may have caricatured the conversation a bit (but not that much!), the point is obvious. Adults must let children be true partners in their interactions and must allow children to influence the directions the conversation takes.[29] The adult also must have a sense of what the child knows and doesn't know and how to expand that knowledge at the right level of detail, which also links to other knowledge the child already has. Our didactic predators fail to truly "partner up" with their child and instead seem more like vintners trying to pour their precious wines of knowledge into the empty bottles they envision as their children's heads. There is also no sense of playfulness here even though "guided play" can be highly effective in engaging children in deep learning.[30]

When all goes well and parents do connect with their children in a spirit of shared discovery and knowledge creation rather than knowledge implantation, such activities are rewarding for all participants. But it can be challenging to interact in this way if a single parent is with the child and two younger siblings, is sleep-deprived and exhausted after working shifts at two jobs, is trying to prepare dinner, and is getting everyone ready for bed. We all need to work together on creating such situations for the children with whom we share our lives. We may also inadvertently treat groups of children differently. For example, in contexts in which parents and children discuss science-related topics, parents tend to engage in explanations more frequently with boys than with girls.[31] This kind of difference makes no sense and I doubt that most parents know that they are behaving differently, yet it matters.

To move toward richer more wonder-building conversations with children, think of specific actions we might take to expand on our wonder tool kits. For example, you could keep an oral or written diary of conversations you had with a child over the past day. In particular, you might try to remember how often you asked open-ended questions as opposed to questions that are not open-ended. You might ask yourself which open-ended questions sparked wonder and delight, and which were more didactic and intimidating.

You could ask what percentage of the time you were talking. Here, it might help to occasionally leave your phone on record for thirty minutes and listen to it later. You may be surprised at how different your recall is from reality. You could also consider the content of both participants' speech. If you were conscientious enough to make elaborations, did they trigger more elaborations from the child as well? Were both of you working together as partners in wonder, or was one of you controlling the agenda? Did you learn anything or have any new questions come to mind that you wanted to look into further? If not, you might ask what derailed your own role as a wondering co-learner. Some adults may find it helpful to set specific goals for their next conversation, such as doubling the percentage of open-ended questions, if they seem low, or asking at least one why or how question about a topic where you have minimal mechanistic understanding yourself. Simply looking at how the diary entries evolve over successive weeks can also provide a sense of how you are progressing.

We shouldn't put all the burden of creating great conversations solely on parents and other adults who may have other important concerns competing for their attention. We can structure the environment so as to gently guide people toward more meaningful interactions. Such situational props can be straightforward. In one approach, longtime collaborators Kathy Hirsh-Pasek and Roberta Golinkoff led a group that inserted signage into grocery stores. The signs posed questions that parents might ask their children in the produce and dairy sections.[32] The signs asked, "Where does milk come from?" and "What's your favorite vegetable?" or "Why is milk good to drink?" and "Why are vegetables good to eat?" Low-SES adults and children were almost four times more likely to have conversations when the signs were in place than when they were absent. No effect was found for middle-SES adults and children, possibly because of a "ceiling effect" in the design, namely that those parents were already chatting more with their children and it was difficult to nudge them even higher in that particular setting. My own guess is that the better off parents may have been nudged toward more why questions (not measured in this study) and that other kinds of questions might have elicited more engaged conversations with their children as well.

The supermarket demonstrations don't mean that you should have questions cards arranged all around your home. The few times that I have seen such attempts, they seemed contrived and stilted. But you might survey your house and think about what things in your home or yard might engender particularly wonder-driven and wonder-growing experiences for the entire family. When we discovered the abundant fossils near our house in Upstate

New York, we soon realized that deliberately lingering for a moment or two whenever we passed them led to many more shared learning experiences and seeking out expert sources. Gentle subtle shifts in where you pause and what you are attending to may bring surprising benefits.

At the least, we could create daily checklists to see how many activities on the list we did at least *once* with a child. For example, we might ask the following ten questions of ourselves.

1. Did I ask at least one why question?
2. Did I ask at least one how question?
3. Did I ask at least one question about something where I didn't know the answer?
4. Did I respond with useful information (either supplying mechanism or seeking out sources) to at least one why or how question from the child?
5. Did I reveal my ignorance at least once about something and describe how I might address it?
6. Did I learn at least one new thing about the world arising from our conversations by the end of day (including time to research it after a child was asleep)?
7. Did I at least once engage in explicit conjectures or speculations that I had never made before?
8. Did I supply at least one elaboration to a child's statement?
9. Did I return to a topic discussed on earlier occasions and supply more mechanistic information?
10. Did I learn at least one new thing about the child's knowledge or interests?

Checklists can be especially effective and low-effort ways of tracking our behaviors and may be more sustainable for many people than daily diaries.[33] This list is merely suggestive. Checklists should embody goals that you find appealing and that would support and develop wonder. They should also evolve over time as your interactions develop. Finally, these checklists, like all others suggested here, should never be seen as a way of checking off an unpleasant chore and thereby freeing oneself from repeating the action. Instead, if the items are well selected, they should whet the appetite for more of the same.

Modeling Mindsets

We can also model the right kinds of curiosity and inquiry for our children and others. We can ask why and how questions more often and ask fewer what, where, and how many questions.[34] We can show how to structure such questions into sequences that are fun shared explorations and are not merciless interrogations. By adopting and developing such "mechanistic mindsets" more often, we enrich our lives and those of others. Part of the art of fine-tuning such mindsets is knowing how deep to dive into in each situation. No one can absorb at one go the fully mechanistic complexity underlying most things. They have to be guided through successively deep layers of elaboration, where higher levels help set up the ability to understand the next level below.

Suppose you decide you want to understand how a virus works, how it infects an organism, reproduces, and then infects another victim. A good teacher might start by describing various types of microorganisms and then quickly zero in on viruses and bacteria, explaining that all viruses, but only a few bacteria, need to enter a host's cell to reproduce. After a pause, the teacher might expand on how viruses can make thousands of copies of themselves while inside a cell and then burst out to infect other cells before eventually leaving that organism to infect another. Throughout this description, the teacher would allude to the presence of many causal steps underlying such things as successfully entering a cell, taking over the cell's machinery, reproducing, assembling its parts into a working virus, and then leaving the cell. This amount of mechanistic explanation, with many allusions, might be all that an adult would want to cognitively digest in one session. But even that amount of information can increase the ability to read articles and listen to conversations about viruses. As that first level of explanation becomes more stable and interconnected with other knowledge, learners might want more details, such as the differences between RNA and DNA viruses and what distinctive properties tend to be associated with each and why. After digesting that knowledge for a while and using it in discourse with others, learners might then want more details of how RNA viruses invade cell membranes.

More than a dozen successively richer levels of mechanistic explanation might be traversed before reaching the state-of-the-art knowledge about how one type of virus works. Only then would we also fully realize all that is still unknown. Wondering about a complex system cannot be completed in one cognitive bite. You have to find appropriately sized and structured chunks of information that set you up to acquire the next level of detail. Gifted teachers package information so as to optimize this process, but we can also learn on our own how to approach a new area, get a sense of the different levels of explanation, and unpack them over time. We can guide children in how to do the same thing. They can learn how to pace themselves as they uncover more and more layers of causal structure and start to grasp natural points at which to pause and consolidate what they have learned. This may be one reason why teaching science through a historical lens often works so well; historical progress often pauses for a period at certain levels before diving deeper and can thereby provide pointers to such levels. We can help children in this process by being a bit more explicit about how we explore and learn, pause and consolidate, and then decide to go deeper if we feel comfortable and are still interested in doing so.

We can model arguing to learn and not arguing to win (unless we are trial lawyers). We can show how arguing can be fun for both parties and a way to test and improve ideas. We can develop such skills in young children who are surprisingly well equipped to adopt them and who will have more productive understandings of the nature of truth as a result. Arguing to learn might be conveyed through daily rituals. One of my childhood friends had times during every family dinner when someone, usually the father, would make a provocative or surprising claim and then encourage everyone to challenge the claim to see if he was mistaken. As a frequent guest at such dinners, I quickly learned that these were meant to be fun learning episodes, not opportunities to destroy others in arguments. As it was a big family, both parents were highly attentive to the tone of the discussion, and when older siblings started to veer into argument styles that seemed designed to conquer and crush their little sisters and brothers, the parents would intervene. Arguing in that aggressive manner was no fun in the long run even if it did provide a momentary thrill. Moreover, it deprived the aggressor of the experience of actually learning. This was nimbly illustrated in real time.

Over the course of many visits to my friend's house, I learned that those dinners were the most fun when I listened carefully and even thought ahead a few steps before entering the argument. It was intimidating at first, but it soon became a natural and a hugely informative way of learning how to argue productively and enjoyably. In fairness to my own parents, our dinners had a bit of the same flavor, but there were only three children in our family who were more spread apart in years and my parents tended toward the mission of instilling intellectual humility in us. They'd bring up a topic, invite us to argue, and then quickly overpower us with counterarguments. To their credit, on those rare occasions when we could mount a reasonable counterargument, they would graciously acknowledge its merit.

Arguing to learn entails a willingness, indeed even an eagerness, to have your mind changed not just about trivial things but also about important ones. You might even want to learn something that dramatically changes your understanding at regular intervals. Consider trying to change your understanding of something important at least once a month. Science constantly progresses and makes dramatic twists and turns, and so should we. When we haven't been conceptually startled with some frequency, we risk falling into confirmation biases rather than disruptive discoveries. For the past few years, I have regularly asked myself if I've changed my understanding of something in the past month. If can't think of anything, I deliberately start looking hard at a long-held belief about some phenomenon to see if it holds up under scrutiny. My beliefs rarely survive a close look, and the holes or flaws can cause startling shifts of perspective. If the gaps don't cause a shift, I look somewhere else until I am startled.

For example, I recently learned that large rock formations can move in the same manner as tides. They are influenced by the gravitational pulls of the moon and the sun.[35] This flew in the face of all my naive notions about the stability of land formations and is now triggering a host of follow-up questions and explorations. For example, I've learned that "tidal" cycles can induce microearthquakes and volcanic activity. I've also learned that the more "elastic" mountain ranges gently sway back and forth every six hours or so in synchrony with the moon's pull. The image of huge mountains rhythmically swaying back and forth suddenly makes the Andes seem like a scene out of Disney's *Fantasia*.

In addition to daily checklists, we might consider weekly or monthly checklists of the following sort:

1. Did I change my understanding in a substantial way? This means not just learning new facts or falsehood; it means learning a new causal model or explanation.
2. Did I disagree openly with a peer at least once in an argue-to-learn way that led to a pleasant and mutually informative conversation?

3. Did I disagree openly with a child at least once in an argue-to-learn way that led to a pleasant and mutually informative conversation?
4. Did I encourage a child at least once to disagree with me or doubt what I said as a way of advancing both of our understandings?

Finally, we might go through the same checklists with our significant others. Almost all of the activities on this checklist are so central to learning and wondering together that they would apply with variations to almost any dyad, as well as to larger groups. Engaging in such activities more with peers may also help better prepare us for futures in which children are bigger parts of our lives.

At the Community and School Levels

We saw how placing a few question cards in dairy and vegetable sections of grocery stores can increase conversations between parents and children about science-related themes. That kind of prompting could easily be implemented on a much larger scale.

Embedded Science

Hirsh-Pasek and Golinkoff and colleagues have planned to extend their supermarket approach to what they call "Thinkscapes."[36] To test the idea, the group designed an urban playground with components that presented spatial puzzles on structures and murals depicting situations that invited story creation. The playgrounds increased caregiver interactions around all of these designed components, suggesting a much wider range of similar embedded prompts in many public spaces. We don't yet know whether the same landscape can increase interactions to the same degree across groups or whether we need to tailor each landscape to particular cultural and economic groups to get peak effects. "Bespoke" playgrounds might be ideal solutions, but it will be easier if a standardized enhancement worked well enough in most settings. Unobtrusive physical facilitators of conversations could be gracefully embedded in virtually all areas of public life.

Building on this empirical work, consider the following three projects:

1. Imagine an escalator that is transparent on all sides so as to reveal its inner mechanisms. Parts are color coded in ways that reveal meaningful functional subunits and causal chains. A display could provide information about the number of parts of each component, their material composition,

and the number of times per hour they go through a cycle. These "seecalators" could advance understandings of mechanical complexity, control loops, and synchronous motions, among many other patterns.

2. Imagine a group of trees in a public space. At dusk, faint LED lights in the ground start blinking a pattern indicating paths and directions of communication along the "wood-wide web," that network of fungi tendrils that interlinks all the trees across species. A cell phone app might enable viewers to shift between LED patterns representing networks that provide nourishment and those that warn other trees to mount an immune response to invaders. If putting LED lights in the ground is too invasive, an app could enable users to view the meadow with the pathways overlaid.
3. Imagine that every dog bowl sold by a major vender had a picture of dog food with fruit slices on top and the following question: "Why must humans eat fruit but not dogs?" This question prompts many follow-up investigations. You might first learn that humans can't synthesize vitamin C (ascorbic acid) out of other foods, but dogs and most other animals can. Humans, primates, guinea pigs, bats, and a few fish cannot. Most of these vitamin C–dependent animals are missing the last step of a longer synthesis chain. Why was this last step selected out in evolution? It remains a puzzling mystery, but one that can be easily grasped by pet owners of all ages.

These examples are merely clumsy placeholders for much better displays created by others. Eight design principles might provide useful guides:

1. Be frugal. Engaging displays will only propagate when they are made as inexpensive as possible.
2. Allow reproduction by others.
3. Situate displays in everyday venues. These include bus stops, small parks, street corners, and even businesses. Do not put them in formal exhibit spaces such as science museums or zoos.
4. Use sparse language and assume little relevant prior knowledge.
5. Pose a puzzle that people would find intriguing and relevant.
6. Provide part of an answer or strong hints and guides to getting fuller answers.
7. Encourage discussions across ages.
8. Use apps for a larger region that record and share ideas about all such displays.

Space, Time, and Other Dimensional Mappings

Some communities have set up scale models of the solar system. A beautiful model system was created in Ithaca, New York, in 1997 in memory of the scientist and science writer Carl Sagan. There are well over two dozen semi-permanent scale models of the solar system in the United States and over one hundred worldwide. Walking the full length of one of these models is an eye opener. The tiny size of the planets with respect to the sun is surprising. A one-meter diameter sun is scaled to an earth the size of a pebble and a Jupiter the size of a grape. Even more startling are the distances between each tiny planet. While the earth is about thirty-three meters away from the sun, which at first seems very far, Pluto is almost a mile away, with a large gap (the asteroid belt) between Mars and Jupiter.

Everyone should have the opportunity to be dazzled by such scales and their grandeur. New questions and wonders about causal influences across great distances can easily emerge. Many other spatial possibilities exist beyond the solar system. The scale of biological structures is fascinating. If a typical human blood cell was the same size as a baseball, an adult would be about ten miles tall. In contrast, a typical virus would be at best a barely visible speck of dust. Murals of this sort would help people understand just how many viruses can be packed into a single cell and perhaps even think about social distancing a bit differently.

Many relative spatial relations could be depicted in built and natural environments for topics as diverse as parts of an animal, a galaxy, or an electronic circuit. We could develop phone apps that could give drivers and passengers senses of scale while driving about. If you are commuting from home to work by the same route each day, every morning as you start out you could ask the app to tell you when you have driven the size of various objects in the body, setting the entire distance to work at the length of your full body (or any other standard you choose). If it was a ten-mile commute, your phone might tell you immediately that you have passed a virus and a red blood cell. It you were interested in cellular-level structures, you might tell the app to expand the scale so that a small freckle was the size of your entire commute. For a different perspective, you can ask it to assume that a carbon atom was a certain size, say the size of a grain of sand. It could then announce when you have crossed the typical size of other familiar things ranging from pepper grains upward. Other people could contribute interesting new spatial progressions to the app, ranging from hiking distances along the Appalachian Trail, to historic ocean voyages, to size difference in new areas of nanotechnology. In addition, the app could give the same relational information on local hiking and bike trails, on longer family car trips, or on transcontinental train trips. If the real distance is long, an app could narrate two continua at the same time so that listeners could get a better sense of relative distances between elements. I can easily imagine a library of thousands of such spatial progressions that could soon be available.

An even more intriguing use of spatial extent would model other continua such as time. For example, imagine a drive from the Boston Public Library to the New York Public Library. Boston is the first appearance of any form of life on earth, and New York is the first appearance of humans. Many drivers would be stunned to learn that the first cells with nuclei appeared around New London, Connecticut, and that multicellular organisms did emerge not until New Haven. Dinosaurs pop up around White Plains and suddenly disappear in Mamaroneck. Humans appear at Grand Central Terminal, just a few blocks from the New York Public Library (see figure 9.2).

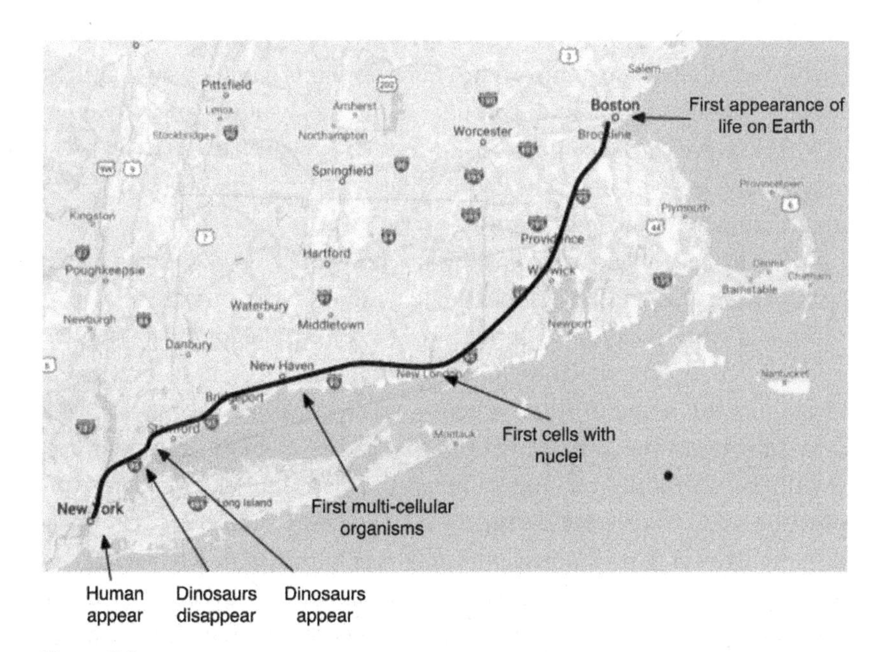

Figure 9.2
Using familiar routes to visualize relations on other spatial, temporal, and causal dimensions. This example maps a trip from Boston to New York onto the timeline from the first life on earth to the emergence of humans.

Similar spatial versions of timelines overlaid on navigation routes could be created for the history of events leading up to a discovery or current version of a modern device, such as the history of indoor lighting starting with the simple flame, to the candle, to oil and gas lamps, to the incandescent light, to LED lighting; or from the first cameras to today's digital versions; or from the first rocket to the latest space vehicle. Topics in developmental psychology and developmental biology could be revealed in the same way. For example, if your home was birth and the office was five years, one might want to hear examples of typical children's vocabularies and sentence structures at certain distances corresponding to months and years. In developmental biology, one could set the scale from the moment of conception to the moment of birth across various organisms.

Any interesting historical period is also easily modeled, ranging from the history of the Vietnam War to the rise and fall of Rome. In many cases, the content of high school– and college-level courses could be integrated with timeline/route mappings. An unlimited set of timelines is possible. One could even show how several different lines of historical discoveries had to co-occur and then converge to yield the final product or how a discovery led to later branching events. Such accounts could easily highlight the cross-cultural and multidisciplinary nature of precursors to discoveries. They could also show the tragic just misses that might have avoided a war or a famine. In some apps, the cheery voice of a favorite scientist or historian could narrate each new significant waypoint.

Other insights might be gained by speeding up or slowing down causal events as revealed on spatial scales. Imagine speeding up mitosis by a factor of ten and corresponding descriptions of each stage at appropriately scaled distances, or slowing down the times of various stages from when a visual image hits the retina to when a person has a conscious recognition of a face. Time is not the only continuum that can be mapped onto distances. Countless other scales can also be mapped. Relative salaries in a large corporation could be revealed. As the lowest wage workers in a large company drove to work at that company, they might be surprised to see that, if their starting point was the CEO's salary and the endpoint was no salary at all, they didn't pass their salary until the last hundred meters of their drive. The total number of moving parts of various devices could be revealed to provide a crude sense of relative complexity.

Causal cycles could be illustrated by using ring roads that circle cities throughout the globe. For example, strategically chosen exits and entrances for I-435 around Kansas City might be used to illustrate the Krebs cycle. Causal chains with discrete steps might be modeled on roads that hop across islands, such as on the Overseas Highway from Miami to Key West, Florida. Other routes with converging and diverging branches might illustrate systems with common causes or common effects.

We could all start doing this today. Between real 3D models and apps that scale information with distances, we could experience interesting patterns not just in the sciences but in any area where there is structure that can be modeled. Not only would these interdisciplinary mappings provide opportunities to understand relative magnitudes but they could foster conversations among cotravelers. Controversial timelines for a topic could both be revealed and invite discussions of which seemed more plausible and why.

Community Events

At the community and school level, we can work toward practices and policies that promote greater engagement with science and engineering. We can have celebratory "egg drop" days where all members of the community are invited to develop vehicles for eggs that will enable them to land intact after being dropped from heights such as twenty meters. Additional categories are often added, such as designing a vehicle that carries an egg the fastest and closest to a target while still not breaking it, or one that has the longest air time without using lighter-than-air assists. Yet another variation is the "naked egg" drop in which contestants design a container to catch a plain egg dropped from a height. These kinds of challenges typically are not tense competitions but rather are playful shared experiences of different designs and their consequences with goofy prizes for all. Having attended several, I witnessed many discussions among the audience and entrants, ad hoc collaborations and innovations, and lots of laughter. These kinds of events invite on-the-spot problem-solving and innovation and insightful suggestions on how failed entries might have survived. Kristi and I organized egg drops around Easter in a Yale residential college where we were heads for eleven years. Many students whom we had recently seen anguishing over a problem set were delighting in the outlandish and clever designs that they and others had made.

Maker Movements

"Makerspaces" have sprouted up in towns and cities throughout the world and have been widely embraced by many education researchers for populations ranging from pre-K to college. The "maker movement" is meant to describe communities who build and create things and share their knowledge with each other either in the same physical makerspace or in online communities. Sharing skills and knowledge is a central theme as is equal access to needed tools and technologies to solve problems. Computational and high-tech tools such as 3D printers are often included, but, in contrast to earlier computer hacker spaces, there is an emphasis on the physical product. Products can range from robots to woven carpets. The movement has been optimistically portrayed as a way to enable anyone to "change the world" when they are given access to a well-equipped makerspace.[37]

The romantic view of massively disruptive technologies emerging out of a makerspace has not yet happened, but that isn't the biggest benefit. Working in these spaces does give participants a better sense of what is involved in fabricating a wide range of objects. It also provides a clearer understanding of how people design and build physical solutions to specific problems, such as the least expensive durable pedestrian bridge that could cross a four-foot stream with as many as four people on it at the same time, or a device that moves through air ducts of a building finding pockets of mold. As part of those design challenges, people start to learn about certain components or mechanisms that repeatedly occur in everyday objects. These include how to convert circular motions to linear ones, how to increase forces with pulleys and gears, how to program a timing circuit, or how to reduce vibration. As makers develop reliable intuitions of how solutions to these problems are commonly employed, they are better equipped to understand, or figure out, how other things work.

In educational settings, makerspaces are also described as democratizing a skill that might have been limited to more privileged children. Spaces can provide new senses of community and belongingness. Makerspaces also embrace creativity, which might not be easily expressed in some classrooms.[38] Many schools now have dedicated makerspaces that help students better understand diverse design and fabrication processes as well as providing insights into recurring ways to improve or add a function to a new device. One effective teaching technique poses a well-thought-out design challenge and supplies tools and materials to meet that challenge. In many cases, however, the main benefits may simply be enjoying the act of creating novel physical things and sharing that activity with others. It is much more difficult to show substantial gains in understanding of science or technology.[39] On the negative side, these makerspaces can be hijacked by companies peddling expensive maker kits for specific challenges. These can certainly be appealing to the overworked teacher, but they often reduce the feeling of being involved in a unique project of creation and synthesis with local materials.

Community makerspaces do seem to help people of all ages rediscover the joy of making things "by hand" and having the satisfaction of solving a problem. People come to better appreciate all that goes into designing and making even the simplest of things. However, they do not have a primary mission of helping others see more clearly how things around them work in both the engineered and natural worlds. Some insights along those lines may come as side benefits, but they are not the central goal.

Mind Change Challenges

To allow our understandings to grow, we need to be continuously open to new ideas and eager to change our minds when appropriate. People do not easily change important core beliefs, but they can if given the right reasons. Intellectual humility involves not only admitting you were mistaken but also feeling a "conceptual change hunger" if you have gone too long without having had an interesting conceptual insight. We might promote an interest in self-driven conceptual change with a web-enabled game that announced the greatest mind-changing idea of the day, week, month, and year. People could contribute cases in which they have changed their minds and ask others to vote on whether they too changed their minds in a similar way. All entries might have a standard format, such as beginning with a why or how question, followed by a 50- to 100-word description of what a person used to think and a 50- to 100-word description of what they now think and why. Others could join in with their own supporting or disagreeing comments or new alternative explanations. If several examples were generated by a group of volunteers through extended discussions, and then posted, those examples could seed a large online community.

Public Performances

I've argued that historical narratives about advances in science and engineering increase understanding of phenomena and devices as well as of the nature of science. If someone from a town had made a major discovery in the distant past or if the town's economy depended on a particular discovery or invention, the town could celebrate that discovery/invention by a historical reenactment of the events leading up to the discovery. If the discovery took generations to come together, that could be reenacted as well along with costume and set changes for each historical epoch. A few science teachers have created such reenactments in their high school classes, and these performances can become some of the strongest memories that students have of high school science content. At the same performances, intermissions could include songs, poems, and limericks created around the same theme. A popular current variation at the graduate level has students dancing their PhD theses.[40] Watching one of the YouTube entries in which a smooth-moving candidate tangos their latest discovery about quantum entanglement is indeed entangling.[41] All these performances do not have to occur in universities and high schools. They can happen in towns and cities as parts of celebrations, fairs, and countless other gatherings.

We can sponsor talks and other events that reveal science in all its beauty to others, just as we can support and become more engaged with our science museums to ensure that they are offering the best possible experiences they can. We might bring back versions of the Chautauqua meetings of over a century ago that brought intellectual pleasures to the general public, not just to the privileged few who can afford to attend expensive conferences in elite destinations. We can pose intriguing puzzles in our local media, just as we now encounter crossword puzzles and sudoku. We could, for example, ask how birds with skinny, uninsulated legs can stand outside for hours in subzero weather without getting frostbite or why Teflon sticks to pans when it doesn't stick to anything else, both of which have known answers. Or we can ask why no one can make room temperature superconductors, which I believe still remains a mystery. We can pose "find the imposter puzzles" in which dueling duos present different arguments or explanations for various phenomena and have the audience vote on who is telling the truth. In the most dramatic cases, a skilled imposter might not be given any advance notice of the topic. I have observed some explanation improv performances along these lines that can be alarmingly seductive. Doing all this with live performances that invite audience participation can result in greater engagement and learning. At the least, online videos of talks might be streamed to groups in the same way that some exercise videos, such as for bike spinning, invite live participants.

At the National Level

At the national level we can work for policies and politicians that not only support science research and education but also support legislation requiring groups to provide explanations of what they do or make. We could ask companies who promote a new car or drug as better than prior ones to explain why the new product is an improvement and to provide plausible causally grounded evidence that the public can understand. At the very least, they should describe experiments demonstrating an improvement, while admitting that they are still puzzled as to why. This would be in stark contrast to disgraceful current advertisements that play on misleading surface similarities as "explanations." One encouraging sign at the national level is the recent launch of a new PBS series, *Elinor Wonders Why*. This show, developed by mechanical engineer Jorge Cham and physicist Daniel Whiteson, focuses on the joy that happens when, while learning how something works, you discover how much more you don't know.[42] The more we can illustrate that process in real people and in animations, the better.

Right to Repair

There is already a movement asking manufacturers to make devices and appliances that are more user fixable. Those movements could be enhanced greatly by linking them to other initiatives mentioned earlier. A more mechanistically literate public could be more empowered consumers. This "right to repair" movement is somewhat closer to increasing mechanistic understanding, but it mostly focuses on rights to have usable repair instructions, reasonably priced and replaceable parts, and specialized repair tools, not on information about how devices such as cell phones actually work. This movement can occur at both the community level and at the national level. It exists as a bike repair shop that I've visited in New Haven that is a blend of maker movement and fix-it culture. It includes groups of farmers across the country who want to be able to fix their expensive farm equipment themselves at a reasonable price. In addition to all these goals, we also have a right to receive simple explanations of what components do and why they are in the device. Some components, such as AI programs that promote addiction to social media web sites, could be real eye-openers when explained.

Science Policy and Funding

The broader public should be meaningfully engaged, wherever possible, in discussions concerning the funding of basic science. They should be able to learn about discoveries and advances that are not simply focused on defense, disease, or deterioration. The potential benefits should include research that helps satisfy wonder regardless of any practical implications. The astonishing changes in cosmological theory over the last few decades may not matter much to our daily lives, but to many outside the field, they are profoundly important steps toward better understanding our ultimate origins and place in the universe. To enable the public to be more engaged requires support from many groups: educators, scientists, science communicators, and politicians. They should join together to ensure that everyone can appreciate the workings of the world in more detail and how this makes life experiences deeper and ever more beautiful. In many cases, such idealistic motivations may end up informing more practical decisions as well, but the public should always understand that a primary reason for going to Mars is not just for economic benefits or national bragging rights. It is simply to enlarge our understanding and appreciation of nature. We must also reject claims that all these proposed ways to inspire and preserve wonder are just meant for an enlightened elite. Children show us otherwise.

Ultimately, greater public engagement with science will make us all better equipped to deal with urgent problems as well. We are all in the midst of the COVID-19 epidemic in which people have died because of inadequate understandings of the virus, of contagion, and of vaccines. With stronger traditions of public engagement with science, we all might have found it easier to behave in the most health-protective ways. Wonder provides far more than a path to more engagement with science and better health. Those who feel able to openly wonder about anything are able to have transformative influences on their societies as a whole. True wondering is without guile or ulterior motive. It instills humility and a need to better understand both what we know and don't know. It can lead to insights that give new rights to the oppressed and steer a planet away from self-destruction. In her book *The Death of Why? The Decline of Questioning and the Future of Democracy*, political writer Andrea Batista Schlesinger maintained that as why questions decline, decreased involvement in critical social and political issues inevitably occurs and threatens democracy itself.

■ ■ ■

In the end, one simple point remains. The very young, and often the very old, have little interest in those material goods that motivate so many people in between. They derive their greatest joys from social connections, from shared experiences, and from discovering new things that amplify those experiences. Young children show us all that we are intrinsically driven to wonder about the world and to address those wonders. When we neglect or even discourage this magnificent cognitive endowment, we are all diminished. But, when we embrace and nourish wonder, we collectively inspire each other and make our interactions with everything and everyone around us more meaningful and fulfilling. Wonder and a love of science are not the exclusive province of an elite few with nerdy tendencies. The gates to the realm of wonder are wide open to young children, and they all enter eagerly and quickly build up skills for learning more about that realm through cycles of exploration and discovery. We only need to keep those gates open and ensure that new insights and delights in the workings of the world continue to expand and be shared throughout our lives.

Notes

Chapter 1

1. For example, Engel, S. (2011). Children's need to know: Curiosity in schools. *Harvard Educational Review, 81*(4), 625–645; Osborne, J., Simon, S., & Collins, S. (2003). Attitudes towards science: A review of the literature and its implications. *International Journal of Science Education, 25*(9), 1049–1079; Oppermann, E., Brunner, M., Eccles, J. S., & Anders, Y. (2018). Uncovering young children's motivational beliefs about learning science. *Journal of Research in Science Teaching, 5*, 399–421; Harter, S. (1981). A new self-report scale of intrinsic versus extrinsic orientation in the classroom: Motivational and informational components. *Developmental Psychology, 17*(3), 300–312; Bonnette, R. N., Crowley, K., & Schunn, C. D. (2019). Falling in love and staying in love with science: Ongoing informal science experiences support fascination for all children, *International Journal of Science Education, 4*, 1626–1643; Potvin, P., & Hasni, A. (2014). Interest, motivation and attitude towards science and technology at K–12 levels: A systematic review of 12 years of educational research. *Studies in Science Education, 50*, 85–129.

2. Tizard, B., & Hughes, M. (1984). *Children learning at home and in school*. Fontana; Sak, R. (2020). Preschoolers' difficult questions and their teachers' responses. *Early Childhood Education Journal, 48*, 59–70.

3. Ronfard, S., Zambrana, I. M., Hermansen, T. K., & Kelemen, D. (2018). Question-asking in childhood: A review of the literature and a framework for understanding its development. *Developmental Review, 49*, 101–120; Butler, L. P., Ronfard, S., & Corriveau, K. H. (Eds.). (2020). *The questioning child: Insights from psychology and education*. Cambridge University Press.

4. Meyer, S. (2019, January 29). Interview with astronaut and TCEA convention keynote Dr. Mae Jemison. TechNotes blog, https://blog.tcea.org/mae-jemison-interview.

5. Donaldson, M. (1979). *Children's minds*. Fontana.

6. Donaldson, M. (1979). The mismatch between school and children's minds. *Human Nature, 2*, 60–67.

7. Thanks to Dick Brodhead for this reference.

8. Daston, L., & Park, K. (1998). *Wonders and the order of nature, 1150–1750*. Zone Books.

9. Sideris, L. H. (2017). *Consecrating science: Wonder, knowledge, and the natural world*. University of California Press.

10. For example, Carson, R., & Lee, K. (1998). *The sense of wonder*. HarperCollins.

11. See also Harrison, P. (2001). Curiosity, forbidden knowledge, and the reformation of natural philosophy in early modern England. *Isis, 9*, 265–290.

12. During the Middle English period, "to marvel" acquired the meaning of "to ask oneself a question in wonder" with specific use of why and how questions (OED 5b), but that sense of marvel is uncommon today.

13. Mead, M. (1932). An investigation of the thought of primitive children, with special reference to animism. *Journal of the Royal Anthropological Institute of Great Britain and Ireland, 62*, 173–190.

14. Simard, S. W., Perry, D. A., Jones, M. D., Myrold, D. D., Durall, D. M., & Molina, R. (1997). Net transfer of carbon between ectomycorrhizal tree species in the field. *Nature, 388*, 579–582; Simard, S. W., & Durall, D. M. (2004). Mycorrhizal networks: A review of their extent, function, and importance. *Canadian Journal of Botany, 82*, 1140–1165.

15. See Dawkins, R. (2000). *Unweaving the rainbow: Science, delusion and the appetite for wonder*. Houghton Mifflin Harcourt.

16. Twain, M. (1883). Two ways of seeing a river. In *Life on the Mississippi*. James R. Osgood.

17. Holmes, R., and an anonymous reviewer both pointed out this much richer characterization of Keats than some have offered; for more on Keats and his medical experiences, see Epstein, J. (1999). The medical Keats. *The Hudson Review, 52*(1), 44–64.

18. Coleridge, S. T. (1895). *Anima poetae: From the unpublished note-books of Samuel Taylor Coleridge*. W. Heinemann.

19. Holmes, R. (2010). *The age of wonder: How the romantic generation discovered the beauty and terror of science*. Vintage, p. 288.

20. Menocal, M. R. (2002). *Culture in the time of tolerance: Al-Andalus as a model for our own time*. Yale Law School; Lowney, C. (2012). *A vanished world: Medieval Spain's golden age of enlightenment*. Simon & Schuster.

21. Starr, S. F. (2013). *Lost enlightenment: Central Asia's golden age from the Arab conquest to Tamerlane*. Princeton University Press.

22. Pinker, S. (2018). *Enlightenment now: The case for reason, science, humanism, and progress*. Penguin.

23. Wissehr, C., Concannon, J., & Barrow, L. H. (2011). Looking back at the Sputnik era and its impact on science education. *School Science and Mathematics, 111*, 368–375; Herold, J. (1974). Sputnik in American education: A history and reappraisal. *McGill Journal of Education/Revue des sciences de l'éducation de McGill, 9*(002).

24. Herold (1974), Sputnik in American education.

25. Leboe, J. P., & Ansons, T. L. (2006). On misattributing good remembering to a happy past: An investigation into the cognitive roots of nostalgia. *Emotion, 6*, 596–610.

26. Harrison (2001), Curiosity, forbidden knowledge, and the reformation of natural philosophy in early modern England; Dolnick, E. (2011). *The clockwork universe: Isaac Newton, the Royal Society, and the birth of the modern world*. HarperCollins.

27. Harrison (2001), Curiosity, forbidden knowledge, and the reformation of natural philosophy in early modern England; see also Daston & Park (1998), *Wonders and the order of nature, 1150–1750*.

28. Dunning, D. (2011). The Dunning–Kruger effect: On being ignorant of one's own ignorance. In J. M. Olson & M. P. Zanna (Eds.), *Advances in social psychology* (vol. 44, pp. 247–296). Academic Press.

29. Hannah Arendt scathingly argued that misdirected wonder about technology and weaponry could isolate humanity from the world we share with all living things: Arendt, H. (2013). *The human condition*. University of Chicago Press.

30. Carson, R. (1998). *Lost woods: The discovered writings of Rachel Carson*. Boston: Beacon Press.

31. Pinker (2018), *Enlightenment now*.

32. Woolfson, P. (2010). The fight over "MACOS": An ideological conflict in Vermont. *Council on Anthropology and Education Quarterly*; Kraus, L. L. (2009). The Fight over MACOS In B. Slater Stern (Ed.) *The new social studies: People, projects and perspectives*. Information Age Publishing, pp. 309–339.

33. Sak (2020), Preschoolers' difficult questions and their teachers' responses.

34. Bloom, P. (2021). *Sweet spot: The pleasures of suffering and the search for meaning*. ECCO.

35. Münch, D., Ezra-Nevo, G., Francisco, A. P., Tastekin, I., & Ribeiro, C. (2020). Nutrient homeostasis—Translating internal states to behavior. *Current Opinion in Neurobiology, 60*, 67–75; Liu, Q., Tabuchi, M., Liu, S., Kodama, L., Horiuchi, W., Daniels, J., . . . Wu, M. N. (2017). Branch-specific plasticity of a bifunctional dopamine circuit encodes protein hunger. *Science, 356*(6337), 534–539.

Chapter 2

1. Summaries of this research are in Keil, F. (2013). *Developmental psychology: The growth of mind and behavior.* W. W. Norton; Goswami, U. (2019). *Cognitive development and cognitive neuroscience: The learning brain.* Routledge; Bjorklund, D. F., & Causey, K. B. (2017). *Children's thinking: Cognitive development and individual differences*. Sage.

2. For example, Turk-Browne, N. B., Jungé, J. A., & Scholl, B. J. (2005). The automaticity of visual statistical learning. *Journal of Experimental Psychology: General, 134,* 552–564.

3. Saffran, J. R., Aslin, R. N., & Newport, E. L. (1996). Statistical learning by 8-month-old infants. *Science, 274*(5294), 1926–1928.

4. Kirkham, N. Z., Slemmer, J. A., & Johnson, S. P. (2002). Visual statistical learning in infancy: Evidence for a domain general learning mechanism. *Cognition; 8,* B35–42; Emberson, L. L., Misyak, J. B., Schwade, J. A., Christiansen, M. H., & Goldstein, M. H. (2019). Comparing statistical learning across perceptual modalities in infancy: An investigation of underlying learning mechanism(s). *Developmental Science,* e12847.

5. Aslin, R. N. (2017). Statistical learning: A powerful mechanism that operates by mere exposure. *Wiley Interdisciplinary Reviews: Cognitive Science,* e1373.

6. Christiansen, M. H. (2019). Implicit statistical learning: A tale of two literatures. *Topics in Cognitive Science, 11*(3), 468–481.

7. Frost, R., Armstrong, B. C., & Christiansen, M. H. (2019). Statistical learning research: A critical review and possible new directions. *Psychological Bulletin, 145*(12), 1128–1153.

8. Cutler, D. M., & Lleras-Muney, A. (2006). *Education and health: Evaluating theories and evidence* (Working Paper No. 12352). National Bureau of Economic Research.

9. Vigen, T. (2015). *Spurious correlations*. Hachette Books.

10. The warning that "correlation does not imply causation" should not be confused with the view that there is no relation between causation and correlation. Reliable correlations normally do arise from stable causal relations, even though those patterns might involve several mediating causal links. Causality in such cases can be inferred through a "causal calculus" that examines the statistical relations between elements and the branching structures in which they are embedded. See Pearl, J., & Mackenzie, D. (2018). *The book of why: the new science of cause and effect.* Basic books.

11. Michotte, A. (1963). *The perception of causality* (T. Miles & E. Miles, Trans.). Methuen. (Original French ed. published 1946).

12. Leslie, A. M. (1984). Spatiotemporal continuity and the perception of causality in infants. *Perception, 13,* 287–305; Leslie, A. M., & Keeble, S. (1987). Do six-month-old infants perceive causality? *Cognition, 25,* 265–288.

13. Kominsky, J. F., Strickland, B., Wertz, A. E., Elsner, C., Wynn, K., & Keil, F. C. (2017). Categories and constraints in causal perception. *Psychological Science, 28*, 1649–1662.

14. Rakison, D. H., & Poulin-Dubois, D. (2001). Developmental origin of the animate–inanimate distinction. *Psychological Bulletin, 127*, 209–228; Opfer, J. E., & Gelman, S. A. (2011). Development of the animate-inanimate distinction. In G. Goswami (Ed.), *The Wiley-Blackwell handbook of childhood cognitive development* (2nd ed., pp. 213–238). Wiley-Blackwell.

15. Woodward, A. L. (1998). Infants selectively encode the goal object of an actor's reach. *Cognition, 69*, 1–34.

16. Johnson, S., Slaughter, V., & Carey, S. (1998). Whose gaze will infants follow? The elicitation of gaze-following in 12-month-olds. *Developmental Science, 1*, 233–238.

17. Heider, F., & Simmel, M. (1944). An experimental study of apparent behavior. *The American Journal of Psychology, 57*(2), 243–259; Gao, T., Newman, G. E., & Scholl, B. J. (2009). The psychophysics of chasing: A case study in the perception of animacy. *Cognitive Psychology, 59*, 154–179.

18. Newman, G. E., Herrmann, P., Wynn, K., & Keil, F. C. (2008). Biases towards internal features in infants' reasoning about objects. *Cognition, 107*, 420–432.

19. Taborda-Osorio, H., & Cheries, E. W. (2018). Infants' agent individuation: It's what's on the insides that counts. *Cognition, 175*, 11–19; Anderson, N., Meagher, K., Welder, A., & Graham, S. A. (2018). Animacy cues facilitate 10-month-olds' categorization of novel objects with similar insides. *PloS One, 13*(11), e0207800.

20. Wertz, A. E. (2019). How plants shape the mind. *Trends in Cognitive Sciences, 23*, 528–531.

21. Rips, L. J., & Hespos, S. J. (2015). Divisions of the physical world: Concepts of objects and substances. *Psychological Bulletin, 141*, 786–811.

22. Paley, W. (1802). *Natural theology, or evidences of the existence and attributes of the deity collected from the appearances of nature.* Oxford University Press.

23. Newman, G. E., Keil, F. C., Kuhlmeier, V. A., & Wynn, K. (2010). Early understandings of the link between agents and order. *Proceedings of the National Academy of Sciences, 107*, 17140–17145; Keil, F. C., & Newman, G. E. (2015). Order, order everywhere, and only an agent to think: The cognitive compulsion to infer intentional agents. *Mind & Language, 30*, 117–139.

24. Ma, L., & Xu, F. (2013). Preverbal infants infer intentional agents from the perception of regularity. *Developmental Psychology, 49*, 1330–1337; Keil & Newman (2015), Order, order everywhere, and only an agent to think.

25. Ma & Xu (2013), Preverbal infants infer intentional agents from the perception of regularity.

26. Spelke, E. S., Breinlinger, K., Macomber, J., & Jacobson, K. (1992). Origins of knowledge. *Psychological Review, 99*, 605–632.

27. Stahl, A. E., & Feigenson, L. (2015). Observing the unexpected enhances infants' learning and exploration. *Science, 348*(6230), 91–94; Stahl, A. E., & Feigenson, L. (2019). Violations of core knowledge shape early learning. *Topics in Cognitive Science, 11*, 136–153.

28. Schulz, L. E., & Bonawitz, E. B. (2007). Serious fun: Preschoolers engage in more exploratory play when evidence is confounded. *Developmental Psychology, 43*, 1045–1050.

29. Muentener, P., & Bonawitz, E. (2018). The development of causal reasoning. In M. Waldmann (Ed.), *Oxford handbook of causal reasoning* (pp. 677–698). Oxford University Press.

30. Kushnir, T., & Gopnik, A. (2005). Young children infer causal strength from probabilities and interventions. *Psychological Science, 16*, 678–683; Schulz, L., Gopnik, A., & Glymour, C. (2007). Preschool children learn about causal structure from conditional interventions. *Developmental Science, 10*, 322–332.

31. Kominsky, J. F., Gerstenberg, T., Pelz, M., Sheskin, M., Singmann, H., Schulz, L., & Keil, F. C. (2021). The trajectory of counterfactual simulation in development. *Developmental Psychology, 57*, 253-268.

32. Nyhout, A., & Ganea, P. A. (2019). The development of the counterfactual imagination. *Child Development Perspectives, 13*, 254–259.

33. Lovejoy, A. O. (1911). The meaning of vitalism. *Science, 33*, 610–614.

34. Driesch, H. (1914). *The history and theory of vitalism.* Macmillan.

35. Inagaki, K., & Hatano, G. (2004). Vitalistic causality in young children's naive biology. *Trends in Cognitive Science, 8*, 356–362.

36. Morris, S. C., Taplin, J. E., & Gelman, S. A. (2000). Vitalism in naive biological thinking. *Developmental Psychology, 36*, 582–595.

37. Carey, S. (1985). *Conceptual change in childhood.* MIT Press.

38. Springer, K., & Keil, F. C. (1989). On the development of biologically specific beliefs: The case of inheritance. *Child Development, 60*, 637–648: Springer, K., & Keil, F. C. (1991). Early differentiation of causal mechanisms appropriate to biological and nonbiological kinds. *Child Development, 62*, 767–781; Gutheil, G., Vera, A., & Keil, F. C. (1998). Do houseflies think? Patterns of induction and biological beliefs in development. *Cognition, 66*, 33–49; Erickson, J. E., Keil, F. C., & Lockhart, K. L. (2010). Sensing the coherence of biology in contrast to psychology: Young children's use of causal relations to distinguish two foundational domains. *Child Development, 81*, 390–409; Lockhart, K. L., & Keil, F. C. (2019). What heals and why? *Monographs of the Society for Research in Child Development, 83*(2); Coley, J. D. (1995). Emerging differentiation of

folkbiology and folkpsychology: Attributions of biological and psychological properties to living things. *Child Development, 66*, 1856–1874.

39. Gelman, S. A. (2011). When worlds collide—Or do they? Implications of explanatory coexistence for conceptual development and change. *Human Development, 54*, 185–190; Legare, C. H., & Shtulman, A. (2018). Explanatory pluralism across cultures and development. In J. Proust & M. Fortier (Eds.), *Metacognitive diversity: An interdisciplinary approach* (pp. 415–432). Oxford University Press.

40. Lockhart & Keil (2019), What heals and why?

Chapter 3

1. Gelfert, A. (2011). Expertise, argumentation, and the end of inquiry. *Argumentation, 25*(3), 297.

2. Harris, P. L. (2002). What do children learn from testimony? In P. Carruthers, M. Siegal, & S. Stich (Eds.), *Cognitive bases of science* (pp. 316–334). Cambridge University Press.

3. See, for example, studies described in: Gelman, S. A. (2009). Learning from others: Children's construction of concepts. *Annual review of psychology, 60*, 115–140; Harris, P. L. (2012). *Trusting what you're told: How children learn from others*. Harvard University Press; Moll, H. (2020). How young children learn from others. *Journal of Philosophy of Education, 54*, 340–355; Gweon, H. (in press). Cognitive foundations of distinctively human social learning and teaching. *Trends in Cognitive Sciences.*

4. In a remarkable recent study, a cognitively diverse group of eight-year-old Australian children were taught why certain materials have specific properties. After a few weekly sessions, the children could explain patterns in the periodic table in terms of electron shell structures and could predict bonding steps between novel configurations of atoms and molecules. This knowledge was still present a year later. Haeusler, C., & Donovan, J. (2020). Challenging the science curriculum paradigm: Teaching primary children atomic-molecular theory. *Research in Science Education, 50*, 23–52.

5. Rogoff, B., Moore, L., Najafi, B., Dexter, A., Correa-Chávez, M., & Solís, J. (2007). Children's development of cultural repertoires through participation in everyday routines and practices. In J. E. Grusec & P. D. Hastings (Eds.), *Handbook of socialization: Theory and research* (pp. 490–515). Guilford Press.

6. Bandura, A. (1965). Behavioral modification through modeling procedures. In L. Krasner, A. Bandura, & L. P. Ullman (Eds.), *Research in behavior modification* (pp. 310–340). Holt, Rinehart & Winston.

7. Horner, V., & Whiten, A. (2005). Causal knowledge and imitation/emulation switching in chimpanzees (*Pan troglodytes*) and children (*Homo sapiens*). *Animal Cognition, 8*, 164–181; Lyons, D. E., Young, A. G., & Keil, F. C. (2007). The hidden

structure of overimitation. *Proceedings of the National Academy of Sciences, 104*, 19751–19756; McGuigan, N., Whiten, A., Flynn, E., & Horner, V. (2007). Imitation of causally opaque versus causally transparent tool use by 3- and 5-year-old children. *Cognitive Development, 22*, 353–364.

8. Johnston, A. M., Holden, P. C., & Santos, L. R. (2017). Exploring the evolutionary origins of overimitation: A comparison across domesticated and non-domesticated canids. *Developmental Science, 20*, e12460; Horner & Whiten (2005), Causal knowledge and imitation/emulation switching.

9. Lyons, D. E., Damrosch, D. H., Lin, J. K., Macris, D. M., & Keil, F. C. (2011). The scope and limits of overimitation in the transmission of artefact culture. *Philosophical Transactions of the Royal Society B: Biological Sciences, 366*(1567), 1158–1167.

10. Hoehl, S., Keupp, S., Schleihauf, H., McGuigan, N., Buttelmann, D., & Whiten, A. (2019). 'Over-imitation': A review and appraisal of a decade of research. *Developmental Review, 51*, 90–108.

11. Lockhart, K. L., Abrahams, B., & Osherson, D. N. (1977). Children's understanding of uniformity in the environment. *Child Development,* 1521–1531; Noyes, A., Keil, F. C., & Dunham, Y. (2020). Institutional actors: Children's emerging beliefs about the causal structure of social roles. *Developmental Psychology, 56*(1), 70–80.

12. Bonawitz, E., Shafto, P., Gweon, H., Goodman, N. D., Spelke, E., & Schulz, L. (2011). The double-edged sword of pedagogy: Instruction limits spontaneous exploration and discovery. *Cognition, 120*(3), 322–330.

13. Callanan, M. A., & Oakes, L. M. (1992). Preschoolers' questions and parents' explanations: Causal thinking in everyday activity. *Cognitive Development, 7*(2), 213–233; Chouinard, M. M., Harris, P. L., & Maratsos, M. P. (2007). *Children's questions: A mechanism for cognitive development.* Monographs of the Society for Research in Child Development. John Wiley; Frazier, B. N., Gelman, S. A., & Wellman, H. M. (2009). Preschoolers' search for explanatory information within adult–child conversation. *Child Development, 80*, 1592–1611; Ronfard, S., Zambrana, I. M., Hermansen, T. K., & Kelemen, D. (2018). Question-asking in childhood: A review of the literature and a framework for understanding its development. *Developmental Review, 49*, 101–120.

14. Liquin, E. G., & Lombrozo, T. (2020). A functional approach to explanation-seeking curiosity. *Cognitive Psychology, 119*, 101276.

15. Chi, M. T., De Leeuw, N., Chiu, M. H., & LaVancher, C. (1994). Eliciting self-explanations improves understanding. *Cognitive Science, 18*, 439–477.

16. Engel, S. (2009). Is curiosity vanishing? *Journal of the American Academy of Child & Adolescent Psychiatry, 48*, 777–779; Engel, S. (2011). Children's need to know: Curiosity in schools. *Harvard Educational Review, 81*(4), 625–645.

17. Tzard, B., & Hughes, M. (1984). *Young children learning*. Harvard University Press.

18. Kurkul, K. E., & Corriveau, K. H. (2018). Question, explanation, follow-up: A mechanism for learning from others? *Child Development, 89*, 280–294.

19. Rohwer, M., Kloo, D., & Perner, J. (2012). Escape from metaignorance: How children develop an understanding of their own lack of knowledge. *Child Development, 83*, 1869–1883; Schneider, W. (2008). The development of metacognitive knowledge in children and adolescents: Major trends and implications for education. *Mind, Brain and Education, 2*, 114–121.

20. Flavell, J. H., Friedrichs, A. G., & Hoyt, J. D. (1970). Developmental changes in memorization processes. *Cognitive Psychology, 1*, 324–340; Yussen, S. R., & Levy, V. M., Jr. (1975). Developmental changes in predicting one's own span of short-term memory. *Journal of Experimental Child Psychology, 19*, 502–508.

21. Taylor, M., Esbensen, B. M., & Bennett, R. T. (1994). Children's understanding of knowledge acquisition: The tendency for children to report that they have always known what they have just learned. *Child Development, 65*, 1581–1604.

22. Cimpian, A. (2016). The privileged status of category representations in early development. *Child Development Perspectives, 10*, 99–104.

23. Similar effects occur in adults: Sloman, S. A., & Rabb, N. (2016). Your understanding is my understanding: Evidence for a community of knowledge. *Psychological Science, 27*, 1451–1460.

24. Richardson, E. D., Sheskin, M., & Keil, F. C. (2021). Scaffolding induces an illusion of self-sufficiency in learners but not observers. *Child Development*. Online version of record before inclusion in an issue: https://doi.org/10.1111/cdev.13506.

25. Hampton, R. R. (2001). Rhesus monkeys know when they remember. *Proceedings of the National Academy of Sciences, 98*, 5359–5362.

26. Goupil, L., Romand-Monnier, M., & Kouider, S. (2016). Infants ask for help when they know they don't know. *Proceedings of the National Academy of Sciences, 113*, 3492–3496.

27. Geurten, M., & Bastin, C. (2019). Behaviors speak louder than explicit reports: Implicit metacognition in 2.5-year-old children. *Developmental Science, 22*, e12742.

28. Lockhart, K. L., Goddu, M. K., Smith, E. D., & Keil, F. C. (2016). What could you really learn on your own? Understanding the epistemic limitations of knowledge acquisition. *Child Development, 87*, 477–493.

29. Shtulman, A., & Carey, S. (2007). Improbable or impossible? How children reason about the possibility of extraordinary events. *Child Development, 78*, 1015–1032.

30. Kominsky, J. F., Zamm, A. P., & Keil, F. C. (2018). Knowing when help is needed: A developing sense of causal complexity. *Cognitive Science, 42*, 491–523.

31. Lockhart, K. L., Goddu, M. K., & Keil, F. C. (2017). Overoptimism about future knowledge: Early arrogance? *Positive Psychology, 12,* 36–46.

32. Lockhart et al. (2017), Overoptimism about future knowledge; Lockhart, K. L., Nakashima, N., Inagaki, K., & Keil, F. C. (2008). From ugly duckling to swan? Japanese and American beliefs about the stability and origins of traits. *Cognitive Development, 23,* 155–179.

33. Harris, P. L. (2012). *Trusting what you're told: How children learn from others.* Harvard University Press.

34. Chen, E. E., Corriveau, K. H., & Harris, P. L. (2013). Children trust a consensus composed of outgroup members—But do not retain that trust. *Child Development, 84,* 269–282; Morgan, T.J.H., Laland, K. N., & Harris, P. L. (2015). The development of adaptive conformity in young children: Effects of uncertainty and consensus. *Developmental Science, 18,* 511–524.

35. Lane, J. D., Wellman, H. M., & Evans, E. M. (2014). Approaching an understanding of omniscience from the preschool years to early adulthood. *Developmental Psychology, 50,* 2380–2392.

36. Johnston, A. M., Sheskin, M., & Keil, F. C. (2019). Learning the relevance of relevance and the trouble with truth: Evaluating explanatory relevance across childhood. *Journal of Cognition and Development, 20,* 555–572.

37. Mills, C. M., & Keil, F. C. (2005). The development of cynicism. *Psychological Science, 16,* 385–390.

38. Mills, C. M., & Grant, M. G. (2009). Biased decision-making: Developing an understanding of how positive and negative relationships may skew judgments. *Developmental Science, 12,* 784–797; Mills, C. M., & Keil, F. C. (2008). Children's developing notions of (im)partiality. *Cognition. 107,* 528–551.

39. Lockhart, K. L., Goddu, M. K., & Keil, F. C. (2018). When saying "I'm best" is benign: Developmental shifts in perceptions of boasting. *Developmental Psychology, 54,* 521–535.

40. Kushnir, T., Vredenburgh, C., & Schneider, L. A. (2013). "Who can help me fix this toy?" The distinction between causal knowledge and word knowledge guides preschoolers' selective requests for information. *Developmental Psychology, 49,* 446–453.

41. Corriveau, K. H., Kim, A. L., Schwalen, C. E., & Harris, P. L. (2009). Abraham Lincoln and Harry Potter: Children's differentiation between historical and fantasy characters. *Cognition, 113,* 213–225.

42. Gweon, H., Pelton, H., Konopka, J. A., & Schulz, L. E. (2014). Sins of omission: Children selectively explore when teachers are under-informative. *Cognition, 132,* 335–341.

43. Harris, P. L., Koenig, M. A., Corriveau, K. H., & Jaswal, V. K. (2018). Cognitive foundations of learning from testimony. *Annual Review of Psychology, 69*, 251–273.

44. Clegg, J. M., Kurkul, K. E., & Corriveau, K. H. (2019). Trust me, I'm a competent expert: Developmental differences in children's use of an expert's explanation quality to infer trustworthiness. *Journal of Experimental Child Psychology, 188*, 10467; Domberg, A., Köymen, B., & Tomasello, M. (2019). Children choose to reason with partners who submit to reason. *Cognitive Development, 52,* 100824.

45. Baum, L. A., Danovitch, J. H., & Keil, F. C. (2008). Children's sensitivity to circular explanations. *Journal of Experimental Child Psychology, 100*, 146–155.

46. See also Mills, C. M., Sands, K. R., Rowles, S. P., & Campbell, I. L. (2019). "I want to know more!" Children are sensitive to explanation quality when exploring new information. *Cognitive Science, 43*, e12706.

47. Mercier, H., Bernard, S., & Clément, F. (2014). Early sensitivity to arguments: How preschoolers weight circular arguments. *Journal of Experimental Child Psychology, 125*, 102–109; Corriveau, K. H., & Kurkul, K. E. (2014). "Why does rain fall?" Children prefer to learn from an informant who uses noncircular explanations. *Child Development, 85*, 1827–1835; Castelain, T., Bernard, S., & Mercier, H. (2018). Evidence that two-year-old children are sensitive to information presented in arguments. *Infancy, 23*, 124–135.

48. Bonawitz, E. B., & Lombrozo, T. (2012). Occam's rattle: Children's use of simplicity and probability to constrain inference. *Developmental Psychology, 48*, 1156–1164.

49. Johnson, S. G., Valenti, J. J., & Keil, F. C. (2019). Simplicity and complexity preferences in causal explanation: An opponent heuristic account. *Cognitive Psychology, 113*, 101222.

50. Koenig, M. A., Cole, C. A., Meyer, M., Ridge, K. E., Kushnir, T., & Gelman, S. A. (2015). Reasoning about knowledge: Children's evaluations of generality and verifiability. *Cognitive Psychology, 83*, 22–39.

51. Johnston, A. M., Sheskin, M., Johnson, S.G.B., & Keil, F. C. (2018). Preferences for explanation generality develop early in biology, but not physics, *Child Development, 8*, 1110–1119.

52. Mills, C. M. (2013). Knowing when to doubt: Developing a critical stance when learning from others. *Developmental Psychology, 49*, 404–418.

53. Kinzler, K. D., Corriveau, K. H., & Harris, P. L. (2011). Children's selective trust in native-accented speakers. *Developmental Science, 14*, 106–111; Marno, H., Guellai, B., Vidal, Y., Franzoi, J., Nespor, M., & Mehler, J. (2016). Infants' selectively pay attention to the information they receive from a native speaker of their language. *Frontiers in Psychology*, 7, 1150.

54. Smith, A., & Stewart, D. (1963). *An inquiry into the nature and causes of the wealth of nations* (Vol. 1). Irwin. (Original work published 1776)

55. Kitcher, P. (1990). The division of cognitive labor. *The Journal of Philosophy, 87,* 5–22.

56. Putnam, H. (1974). Meaning and reference. *The Journal of Philosophy, 70,* 699–711.

57. Bromme, R., & Thomm, E. (2016). Knowing who knows: Laypersons' capabilities to judge experts' pertinence for science topics. *Cognitive Science, 40,* 241–252.

58. Lutz, D. J., & Keil, F. C. (2002). Early understanding of the division of cognitive labor. *Child Development, 73,* 1073–1084.

59. Keil, F. C., Stein, C., Webb, L., Billings, V. D., & Rozenblit, L. (2008). Discerning the division of cognitive labor: An emerging understanding of how knowledge is clustered in other minds. *Cognitive Science, 32,* 259–300.

60. Danovitch, J., & Keil, F. C. (2004). Should you ask a fisherman or a biologist? Developmental shifts in ways of clustering knowledge. *Child Development, 75,* 918–931.

61. Danovitch, J. H. (2019). Growing up with Google: How children's understanding and use of internet-based devices relates to cognitive development. *Human Behavior and Emerging Technologies, 1,* 81–90.

62. The word "debate" may have more positive connotations.

63. Stein, N. L., & Miller, C. A. (1991). I win—You lose: The development of argumentative thinking. In J. F. Voss, D. N. Perkins, & J. Segal (Eds.), *Informal reasoning and education* (pp. 265–290). Erlbaum; Pirie, M. (2015). *How to win every argument: The use and abuse of logic.* Erlbaum; Kleiser, G. (1912). *How to argue and win.* Funk & Wagnalls.

64. See Roth, M. S. (2019, August 29). Don't dismiss "safe spaces." *New York Times.*

65. Kuhn, D. (1991). *The skills of argument.* Cambridge University Press.

66. Kuhn, D., & Katz, J. (2009). Are self-explanations always beneficial? *Journal of Experimental Child Psychology, 103,* 386–394; Brem, S. K., & Rips, L. J. (2000). Explanation and evidence in informal argument. *Cognitive Science, 24,* 573–605; Kuhn, D., & Udell, W. (2003). The development of argument skills. *Child Development, 74*(5), 1245–1260.

67. Stein, N. L., & Bernas, R. (1999). The early emergence of argumentative knowledge and skill. In J. Andriessen & P. Corrier (Eds.), *Foundations of argumentative text processing* (pp. 97–116). Amsterdam University Press; Kuczynski, L., & Kochanska, G. (1990). Development of children's noncompliance strategies from toddlerhood to age 5. *Developmental Psychology, 26,* 398–408; Mercier, H. (2011). Reasoning serves argumentation in children. *Cognitive Development, 26,* 177–191.

68. Dunbar, K. (2000). How scientists think in the real world: Implications for science education. *Journal of Applied Developmental Psychology, 21*, 49–58.

69. Andriessen, J. (2006). Arguing to learn. In R. K. Sawyer (Ed.), *The Cambridge handbook of the learning sciences* (pp. 443–460). Cambridge University Press; Asterhan, C. S., & Schwarz, B. B. (2016). Argumentation for learning: Well-trodden paths and unexplored territories. *Educational Psychologist, 51*(2), 164–187; Osborne, J. (2010). Arguing to learn in science: The role of collaborative, critical discourse. *Science, 328*(5977), 463–466.

70. Fisher, M., Knobe, J., Strickland, B., & Keil, F. C. (2017). The influence of social interaction on intuitions of objectivity and subjectivity. *Cognitive Science, 41*, 1119–1134.

71. Stein & Miller (1991), I win—You lose.

72. Domberg, A., Köymen, B., & Tomasello, M. (2019). Children choose to reason with partners who submit to reason. *Cognitive Development, 52*, 100824; Domberg, A., Köymen, B., & Tomasello, M. (2018). Children's reasoning with peers in cooperative and competitive contexts. *British Journal of Developmental Psychology, 36*, 64–77.

73. Asterhan & Schwarz (2016), Argumentation for learning.

74. E.g., Johnson, S. G., Zhang, J., & Keil, F. C. (in press). Win–win denial: The psychological underpinnings of zero-sum thinking. *Journal of Experimental Psychology: General*; Fisher, M., Knobe, J. Strickland, B., & Keil, F. C. (2018). The tribalism of truth. *Scientific American, 318*, 50–53.

Chapter 4

1. Nicholson, D. J. (2012). The concept of mechanism in biology. *Studies in History and Philosophy of Science, Part C: Studies in History and Philosophy of Biological and Biomedical Sciences, 43*, 152–163; Garson, J. (2019). *What biological functions are and why they matter*. Cambridge University Press.

2. USGS. Volcano hazards program: About volcanoes. https://www.usgs.gov/natural-hazards/volcano-hazards/about-volcanoes.

3. See Craver, C. F. (2013). Functions and mechanisms: A perspectivalist view. In P. Huneman (Ed.), *Functions: Selection and mechanisms* (pp. 133–158). Springer; Millikan, R. (1989). In defense of proper functions. *Philosophy of Science, 56*, 288–302.

4. Adapted from Craver, C. F. (2007). *Explaining the brain: Mechanisms and the mosaic unity of neuroscience*. Oxford University Press; and Craver, C., & Tabery, J. (2019, Summer ed.). Mechanisms in science. In E. N. Zalta (Ed.), *Stanford encyclopedia of philosophy*. https://plato.stanford.edu/archives/sum2019/entries/science-mechanisms.

5. Hobbes, T. (2006). *Leviathan: A Critical Edition*. G.A.J. Rogers and Karl Schuhmann (Eds.). Thoemmes. (Original work published 1651).

6. Bechtel, W., & Richardson, R. C. (2010 [1993]). *Discovering complexity: Decomposition and localization as strategies in scientific research* (2nd ed.). MIT Press/Bradford Books; Craver, C. F. (2001). Role functions, mechanisms and hierarchy. *Philosophy of Science, 68*, 31–55; for links between the new mechanists and reductionism, see also Rosenberg, A. (2020). *Reduction and mechanism.* Cambridge University Press.

7. See Glennan, S., & Illari, P. (2017). Varieties of mechanisms. In *The Routledge handbook of mechanisms and mechanical philosophy* (pp. 91–103). Routledge.

8. Lecompte, G. K., & Gratch, G. (1972). Violation of a rule as a method of diagnosing infants' levels of object concept. *Child Development, 43*, 385–396.

9. To get a sense of how a study of this sort might be done correctly, see: Kibbe, M. M., & Leslie, A. M. (2019). Conceptually rich, perceptually sparse: Object representations in 6-month-old infants' working memory. *Psychological Science, 30*, 362–375.

10. Frazier, B. N., Gelman, S. A., & Wellman, H. M. (2009). Preschoolers' search for explanatory information within adult–child conversation. *Child Development, 80*, 1592–1611.

11. Frazier, B. N., Gelman, S. A., & Wellman, H. M. (2016). Young children prefer and remember satisfying explanations. *Journal of Cognition and Development, 17*, 718–736.

12. Chouinard, M. M., Harris, P. L., & Maratsos, M. P. (2007). *Children's questions: A mechanism for cognitive development.* Monographs of the Society for Research in Child Development. John Wiley; Kurkul, K. E., & Corriveau, K. H. (2018). Question, explanation, follow-up: A mechanism for learning from others? *Child Development, 89*, 280–294.; Frazier et al. (2009), Preschoolers' search for explanatory information within adult–child conversation; Frazier et al. (2016), Young children prefer and remember satisfying explanations.

13. Chouinard, Harris, & Maratsos (2007). *Children's questions*; Hickling, A. K., & Wellman, H. M. (2001). The emergence of children's causal explanations and theories: Evidence from everyday conversation. *Developmental Psychology, 37*, 668–683; Frazier et al. (2016), Young children prefer and remember satisfying explanations.

14. Lockhart, K. L., Chuey, A., Kerr, S., & Keil, F. C. (2019). The privileged status of knowing mechanistic information: An early epistemic bias. *Child Development, 90*, 1772–1788.

15. Kelemen, D. (1999). The scope of teleological thinking in preschool children. *Cognition, 70*, 241–272.

16. Joo, S., Yousif, S. R., & Keil, F. C. (2020). Implicit questions shape information preferences. In S. Denison., M. Mack, Y. Xu, & B. C. Armstrong (Eds.), *Proceedings of the 42nd annual conference of the Cognitive Science Society* (pp. 1265–1271). Cognitive Science Society.

17. Chuey, A., Lockhart, K., Sheskin, M., & Keil, F. (2020). Children and adults selectively generalize mechanistic knowledge. *Cognition, 199,* 104231; Trouche, E., Chuey, A., Lockhart, K., & Keil, F. (2017). Why teach how things work? Tracking the evolution of children's intuitions about complexity. In T. T. Rogers, M. Rau, X. Zhu, & C. W. Kalish (Eds.), *Proceedings of the 40th annual conference of the Cognitive Science Society* (p. 1126). Cognitive Science Society.

18. Buchanan, D. W., & Sobel, D. M. (2011). Mechanism-based causal reasoning in young children. *Child Development, 8,* 2053–2066; Kushnir, T., Vredenburgh, C., & Schneider, L. A. (2013). "Who can help me fix this toy?" The distinction between causal knowledge and word knowledge guides preschoolers' selective requests for information. *Developmental Psychology, 49,* 446–453.

19. Mills, C. M., Sands, K. R., Rowles, S. P., & Campbell, I. L. (2019). "I want to know more!" Children are sensitive to explanation quality when exploring new information. *Cognitive Science, 43,* e12706.

20. Lombrozo, T., & Gwynne, N. Z. (2014). Explanation and inference: Mechanistic and functional explanations guide property generalization. *Frontiers in Human Neuroscience, 8,* 700.

21. Lombrozo, T., Kelemen, D., & Zaitchik, D. (2007). Inferring design: Evidence of a preference for teleological explanations in patients with Alzheimer's disease. *Psychological Science, 18,* 999–1006; Kelemen, D., & Rosset, E. (2009). The human function compunction: Teleological explanation in adults. *Cognition, 111,* 138–143.

22. Lombrozo, T., & Wilkenfeld, D. (2019). Mechanistic versus functional understanding. In S. R. Grimm (Ed.), *Varieties of understanding: New perspectives from philosophy, psychology, and theology* (pp. 209–229). Oxford University Press.

23. Kominsky, J. F., Zamm, A. P., & Keil, F. C. (2018). Knowing when help is needed: A developing sense of causal complexity. *Cognitive Science, 42,* 491–523.

24. Ahl, R. E., & Keil, F. C. (2017). Diverse effects, complex causes: Children use information about machines' functional diversity to infer internal complexity. *Child Development, 88,* 828–845.

25. Ahl, R., Amir, D., & Keil, F. C. (2020). The world within: Children are sensitive to internal complexity cues. *Journal of Experimental Child Psychology, 200,* 104932.

26. Erb, C. D., Buchanan, D. W., & Sobel, D. M. (2013). Children's developing understanding of the relation between variable causal efficacy and mechanistic complexity. *Cognition, 129,* 494–500.

27. Trouche et al. (2017), Why teach how things work?

28. Chuey, A., MacCarthy, A., Lockhart, K. L. Trouche, E., Sheskin, M., & Keil, F. C. (2021). No guts, no glory: Underestimating the benefits of providing children with mechanistic details. *npj Science of Learning.*

29. Betz, N., McCarthy, A. M., & Keil, F. C. (2021). Adult intuitions about mechanistic content in elementary school science lessons. In T. Fitch, C. Lamm. H. Leder, & K. Tessmar (Eds.), *Proceedings of the 43rd annual conference of the Cognitive Science Society* (pp. 1353–1359). Cognitive Science Society.

30. Betz, N. & Keil, F. C. (2021). Mechanistic learning goals enhance elementary student understanding and enjoyment of heart lessons. In T. Fitch, C. Lamm. H. Leder, & K. Tessmar (Eds.), *Proceedings of the 43rd annual conference of the Cognitive Science Society* (pp. 2031–2037). Cognitive Science Society.

31. Schwartz, D. L., Bransford, J. D., & Sears, D. (2005). Efficiency and innovation in transfer. In J. P. Mestre, *Transfer of learning from a modern multidisciplinary perspective* (pp. 1–51). Information Age; MacLeod, C. M. (2008). Implicit memory tests: Techniques for reducing conscious intrusion. In J. Dunlosky & R. A. Bjork (Eds.), *Handbook of metamemory and memory* (pp. 245–263). Psychology Press.

32. MacLeod, C. M. (1988). Forgotten but not gone: Savings for pictures and words in long-term memory. *Journal of Experimental Psychology: Learning, Memory, and Cognition, 14,* 195–212; Roediger, H. L. (1990). Implicit memory: Retention without remembering. *American Psychologist, 45,* 1043–1056.

33. Perkins, D., Jay, E., & Tishman, S. (1993). New conceptions of thinking: From ontology to education. *Educational Psychologist, 28,* 67–85.

34. Keil, F. C., Stein, C., Webb, L., Billings, V. D., & Rozenblit, L. (2008). Discerning the division of cognitive labor: An emerging understanding of how knowledge is clustered in other minds. *Cognitive Science, 32,* 259–300.

35. Sobel, D., & Sommerville, J. (2010). The importance of discovery in children's causal learning from interventions. *Frontiers in Psychology, 1,* 176; Schulz, L. E., Gopnik, A., & Glymour, C. (2007). Preschool children learn about causal structure from conditional interventions. *Developmental Science, 10*(3), 322–332; Woodward, J. (2005). *Making things happen: A theory of causal explanation.* Oxford University Press.

36. Wilkins, J., Schoville, B., Brown, K., & Chazan, M. (2012). Evidence for early hafted hunting technology. *Science, 338,* 942–946. https://doi.org/10.1126/science.1227608; Bradfield, J., Lombard, M., Reynard, J., & Wurz, S. (2020). Further evidence for bow hunting and its implications more than 60 000 years ago: Results of a use-trace analysis of the bone point from Klasies River Main site, South Africa. *Quaternary Science Reviews, 236,* 10629; Brown, K. S., Marean, C. W., Jacobs, Z., Schoville, B. J., Oestmo, S., Fisher, E. C., . . . Matthews, T. (2012). An early and enduring advanced technology originating 71,000 years ago in South Africa. *Nature, 491*(7425), 590–593.

37. Haidle proposes that changes in working memory may have been critical to the transition from simple one object rocklike tools to multipart tools: Haidle, S. M. (2010). Working-memory capacity and the evolution of modern cognitive potential:

Implications from animal and early human tool use. *Current Anthropology, 51,* 149–166.

38. Boyd, R., Richerson, P. J., & Henrich, J. (2011). The cultural niche: Why social learning is essential for human adaptation. *Proceedings of the National Academy of Sciences, 108*(Suppl. 2), 10918–10925; Henrich, J. (2017). *The secret of our success: How culture is driving human evolution, domesticating our species, and making us smarter.* Princeton University Press.

39. Danovitch, J. H., & Keil, F. C. (2004). Should you ask a fisherman or a biologist? Developmental shifts in ways of clustering knowledge. *Child Development, 75,* 918–931.

40. Busch, J. T., Watson-Jones, R. E., & Legare, C. H. (2018). Cross-cultural variation in the development of folk ecological reasoning. *Evolution and Human Behavior, 39*(3), 310–319; Bender, A., Beller, S., & Medin, D. L. (2017). Causal cognition and culture. In M. Waldmann (Ed.), *Oxford handbook of causal reasoning* (pp. 717–738). Oxford University Press.

41. Thomas, N., & Nigam, S. (2018). Twentieth-century climate change over Africa: Seasonal hydroclimate trends and Sahara desert expansion. *Journal of Climate, 31*(9), 3349–3370.

42. Terryn, E. (2019). A right to repair? Towards sustainable remedies in consumer law. *European Review of Private Law, 27,* 851–873; Hernandez, R. J., Miranda, C., & Goñi, J. (2020). Empowering sustainable consumption by giving back to consumers the 'right to repair.' *Sustainability, 12,* 850.

43. Tiger Brown, T. (2012). The death of shop class and America's skilled workforce. *Forbes,* May 30, 2012. https://www.forbes.com/sites/tarabrown/2012/05/30/the-death-of-shop-class-and-americas-high-skilled-workforce/?sh=72ad9b88541f.

44. As a nine-year-old, I stumbled upon two fat books entitled *The Boy Mechanic 1 & 2* (1913, 195). The covers of both books depicted a boy building a complex device. Those books led me to build a pitching machine and a land sailer. Neither worked that well, and both were quite dangerous by today's standards, but it was intensely thrilling to take the book's sketchy plans and turn them into reality. If those books had been titled *The Young Mechanic* instead, perhaps far more girls from my generation and earlier ones could have experienced similar thrills.

45. Samek, W., Montavon, G., Vedaldi, A., Hansen, L. K., & Müller, K. R. (2019). Explainable AI—Preface. In W. Samek, G. Montavon, A. Vedaldi, L. K. Hansen, & K. R. Müller (Eds.). *Explainable AI: Interpreting, explaining and visualizing deep learning* (pp. v–vii). Springer.

46. Safire, W. (1986, June 1). On language: Class cleavage. *New York Times,* section 6, p. 12.

Chapter 5

1. Keil, F. (1980). Development of the ability to perceive ambiguities: Evidence for the task specificity of a linguistic skill. *Journal of Psycholinguistic Research, 9*, 219–230.

2. Heafford, M. R. (2016). *Pestalozzi: His thought and its relevance today*. Routledge.

3. Dewey, J. (1910). Concrete and abstract thinking, chapter 10 in *How we think* (pp. 135–144; see p. 133). D.C. Heath.

4. Standing, E. M. (1962). *Maria Montessori: Her life and work*. New American Library; Lillard, A. S. (2016). *Montessori: The science behind the genius*. Oxford University Press; Montessori, M. (1989). *The child, society, and the world: Unpublished speeches and writings* (Vol. 7). Clio; Montessori, M. (1997). *The California lectures of Maria Montessori 1915*. Clio; Montessori, M. (2013). *The 1913 Rome lectures*. Montessori-Pierson; Lillard, A., & Else-Quest, N. (2006). The early years: Evaluating Montessori education. *Science, 313*(5795), 1893–1894.

5. Montessori, M. (1995). *The absorbent mind* (C. A. Claremont, Trans.). Henry Holt. (Original work published 1967).

6. Montessori (1995), *The absorbent mind*.

7. Lillard, A. S. (2019). Shunned and admired: Montessori, self-determination, and a case for radical school reform. *Educational Psychology Review, 31*, 939–965; Lillard, A. S. (2018). Rethinking education: Montessori's approach. *Current Directions in Psychological Science, 27*, 395–400.

8. Montessori, M. (1948). *From childhood to adolescence*. Schocken.

9. Montessori, M. (2012). *The 1946 London lectures*. Montessori-Pierson.

10. Montessori (1995), *The absorbent mind*, 240–241.

11. Montessori (2012), *The 1946 London lectures*.

12. Montessori (1948), *From childhood to adolescence*.

13. Lillard & Else-Quest (2006), The early years.

14. Simons, D. J., & Keil, F. C. (1995). An abstract to concrete shift in the development of biological thought: The insides story. *Cognition, 56*, 129–163.

15. Lupyan, G., & Winter, B. (2018). Language is more abstract than you think, or, why aren't languages more iconic? *Philosophical Transactions of the Royal Society B: Biological Sciences, 373*(1752), 20170137.

16. Quine, W. V. O. (2013). *Word and object*. MIT Press; Markman, E. M. (1990). Constraints children place on word meanings. *Cognitive Science, 14*, 57–77.

17. Ponari, M., Norbury, C. F., & Vigliocco, G. (2018). Acquisition of abstract concepts is influenced by emotional valence. *Developmental Science, 21*, e12549.

18. Fodor, J. A., Garrett, M. F., Walker, E. C., & Parkes, C. H. (1980). Against definitions. *Cognition, 8*, 263–367.

19. Keil, F. C., & Batterman, N. (1984). A characteristic-to-defining shift in the development of word meaning. *Journal of Verbal Learning and Verbal Behavior, 23*, 221–236.

20. Chi, M. T., Feltovich, P. J., & Glaser, R. (1981). Categorization and representation of physics problems by experts and novices. *Cognitive Science, 5*, 121–152.

21. Simons & Keil (1995), An abstract to concrete shift in the development of biological thought.

22. Gelman, S. A. (2003). *The essential child: Origins of essentialism in everyday thought.* Oxford University Press.

23. Carstensen, A., Zhang, J., Heyman, G. D., Fu, G., Lee, K., & Walker, C. M. (2019). Context shapes early diversity in abstract thought. *Proceedings of the National Academy of Sciences, 116*(28), 13891–13896.

24. Uttal, D. H., O'Doherty, K., Newland, R., Hand, L. L., & DeLoache, J. (2009). Dual representation and the linking of concrete and symbolic representations. *Child Development Perspectives, 3*, 156–159.

25. National Research Council. (2007). *Taking science to school: Learning and teaching science in grades K–8*. National Academies Press; Clements, D. H., & Sarama, J. (2016). Math, science, and technology in the early grades. *The Future of Children, 26*, 75–94; Donaldson, M. (1979). The mismatch between school and children's minds. *Human Nature, 2, 60–67.*

26. Gentner, D. (2010). Bootstrapping the mind: Analogical processes and symbol systems. *Cognitive Science, 34*, 752–775. Richland, L. E., Morrison, R. G., & Holyoak, K. J. (2006). Children's development of analogical reasoning: Insights from scene analogy problems. *Journal of Experimental Child Psychology, 94*, 249–273.

27. Walker, C. M., & Gopnik, A. (2014). Toddlers infer higher-order relational principles in causal learning. *Psychological Science, 25*, 161–169; Walker, C. M., Hubachek, S. Q., & Vendetti, M. S. (2018). Achieving abstraction: Generating far analogies promotes relational reasoning in children. *Developmental Psychology, 54*, 1833–1841.

28. Wojcik, J. (1983). Metaphors in children's speech acquisition. *Poetics Today, 4*, 297–307; Chukovsky, K. (1974). *From two to five* (M. Morton, Trans.). University of California Press.

29. Gentner, D., & Stuart, P. (1984). *Metaphor as structure-mapping: What develops.* University of Illinois at Urbana-Champaign, Center for the Study of Reading (Technical Report No. 315).

30. For summaries, see Keil, F. (2013). *Developmental psychology: The growth of mind and behavior.* W. W. Norton; Gelman, R., & Baillargeon, R. (1983). Review of some Piagetian concepts. In J. H. Flavell & E. M. Markman (Eds.), Cognitive development, Vol. 3 of P. H. Mussen (General Ed.), *Handbook of child psychology* (pp. 167–230). New York: Wiley; Donaldson, M. C. (1979). *Children's minds.* Fontana.

31. For example, Blair, C. (2016). Developmental science and executive function. *Current Directions in Psychological Science, 25,* 3–7; Zelazo, P. D. (2015). Executive function: Reflection, iterative reprocessing, complexity, and the developing brain. *Developmental Review, 38,* 55–68.

32. For example, Diamond, A., & Lee, K. (2011). Interventions shown to aid executive function development in children 4 to 12 years old. *Science, 333*(6045), 959–964; Lawson, G. M., Hook, C. J., & Farah, M. J. (2018). A meta-analysis of the relationship between socioeconomic status and executive function performance among children. *Developmental Science, 21,* e12529.

33. Chase, W. G., & Simon, H. A. (1973). Perception in chess. *Cognitive Psychology, 4,* 55–81; Gobet, F., & Simon, H. A. (1998). Expert chess memory: Revisiting the chunking hypothesis. *Memory, 6,* 225–255.

34. Chi, M. T. (1978). Knowledge structures and memory development. *Children's thinking: What develops, 1,* 75–96; Chi, M. T., & Koeske, R. D. (1983). Network representation of a child's dinosaur knowledge. *Developmental Psychology, 19,* 29–39.

35. Keil F. C. (1989). *Concepts, kinds, and cognitive development.* MIT Press; Keil & Batterman (1984), A characteristic-to-defining shift in the development of word meaning.

36. Kuhn, T. S. (2012). *The structure of scientific revolutions.* University of Chicago Press.

37. Keil, F. C. (2010). Conceptual development and change. In P. C. Hogan (Ed.), *The Cambridge encyclopedia of the language sciences* (pp. 197–199). Cambridge University Press.

38. Carey, S. (2009). Where our number concepts come from. *Journal of Philosophy, 106,* 220–254; Spelke, E. S. (2016). Core knowledge. In D. Barner & A. S. Baron (Eds.), *Core knowledge and conceptual change* (pp. 279–300). Oxford University Press.

39. Simons & Keil (1995), An abstract to concrete shift in the development of biological thought.

40. For example, Larkin, D. (2012). Misconceptions about "misconceptions": Preservice secondary science teachers' views on the value and role of student ideas. *Science Education, 96,* 927–959.

41. Klahr, D., & Nigam, M. (2004). The equivalence of learning paths in early science instruction: Effects of direct instruction and discovery learning. *Psychological Science, 15,* 661–667.

42. For example, Schwartz, D., Bransford, J., & Sears, D. (2005). Efficiency and innovation in transfer. In J. Mestre (Ed.), *Transfer of learning from a modern multidisciplinary perspective* (pp. 1–51). Information Age.

43. For example, National Research Council (2007), *Taking science to school*; National Research Council. (2012). *A framework for K–12 science education: Practices, crosscutting concepts, and core ideas*. National Academies Press.

44. Duncan, R. G., & Rivet, A. E. (2013). Science learning progressions. *Science, 339*, 396-39; Duschl, R. (2008). Science education in three-part harmony: Balancing conceptual, epistemic, and social learning goals. *Review of Research in Education, 32*, 268–291; National Research Council (2012), *A framework for K–12 science education.*

45. Guilford Public Schools (2011, July 11). Science curriculum: Kindergarten–grade 10. https://guilfordschools.org/pdf/curriculum/Science%20Curriculum%20June%2013_%202011.pdf.

46. Simon, H. A. (1977). The organization of complex systems. In R. S. Cohen & M. W. Wartofsky (Eds.), *Models of discovery* (pp. 245–261). Springer.

47. Schwartz, D. L., Chase, C. C., Oppezzo, M. A., & Chin, D. B. (2011). Practicing versus inventing with contrasting cases: The effects of telling first on learning and transfer. *Journal of Educational Psychology, 103*, 759–775.

48. Schwartz et al. (2005), Efficiency and innovation in transfer; Schwartz et al. (2011), Practicing versus inventing with contrasting cases.

49. Grotzer, T., & Mittlefehldt, S. (2012). The role of metacognition in students' understanding and transfer of explanatory structures in science. In A. Zohar & Y. J. Dori (Eds.), *Metacognition in science education* (pp. 79–99). Springer.

50. Koerber, S., Mayer, D., Osterhaus, C., Schwippert, K., & Sodian, B. (2015). The development of scientific thinking in elementary school: A comprehensive inventory. *Child Development, 86*, 327–336.

51. Grotzer & Mittlefehldt (2012), The role of metacognition in students' understanding and transfer of explanatory structures in science.

52. See also Dawkins, R. (2000). *Unweaving the rainbow: Science, delusion and the appetite for wonder*. Houghton Mifflin Harcourt.

53. More than fifty years later the following piece appeared in *Science*: E. Stokstad. (2016, May 10). How the Venus flytrap got its taste for meat.

54. Bemm, F., Becker, D., Larisch, C., Kreuzer, I., Escalante-Perez, M., Schulze, W. X., . . . Hedrich, R. (2016). Venus flytrap carnivorous lifestyle builds on herbivore defense strategies. *Genome Research, 26*(6), 812–825.

Chapter 6

1. PBS NewsHour (2018, March 6). Many preschool teachers are scared of teaching STEM: Here's a solution that might help. https://www.pbs.org/newshour/show/many-preschool-teachers-are-scared-of-teaching-stem-heres-a-solution-that-might-help.

2. For example, Harlen, W. (1997). Primary teachers' understanding in science and its impact in the classroom. *Research in Science Education, 27*, 323–337; Beilock, S. L., Gunderson, E. A., Ramirez, G., & Levine, S. C. (2010). Female teachers' math anxiety affects girls' math achievement. *Proceedings of the National Academy of Sciences, 107*, 1860–1863; Archer, L., DeWitt, J., Osborne, J., Dillon, J., Willis, B., & Wong, B. (2012). Science aspirations, capital, and family habitus: How families shape children's engagement and identification with science. *American Educational Research Journal, 49*, 881–908; Michaluk, L., Stoiko, R., Stewart, G., & Stewart, J. (2018). Beliefs and attitudes about science and mathematics in pre-service elementary teachers, STEM, and non-STEM majors in undergraduate physics courses. *Journal of Science Education and Technology, 27*, 99–113.

3. Bursal, M., & Paznokas, L. (2006). Mathematics anxiety and preservice elementary teachers' confidence to teach mathematics and science. *School Science and Mathematics, 106*, 173–180; van Aalderen-Smeets, S. I., Walma van der Molen, J. H., & Asma, L. J. (2012). Primary teachers' attitudes toward science: A new theoretical framework. *Science Education, 96*, 158–182; Zielinski, S. (2010, January 26). Elementary school teachers pass on math fear to girls. Smithsonianmag.com. https://www.smithsonianmag.com/science-nature/elementary-school-teachers-pass-on-math-fear-to-girls-22912908/#IGvUJ3GzFu8s8d8y.99.

4. Bathgate, M. E., Schunn, C. D., & Correnti, R. (2014). Children's motivation toward science across contexts, manner of interaction, and topic. *Science Education, 98*, 189–215.

5. Johnston, J., & Ahtee, M. (2006). Comparing primary student teachers' attitudes, subject knowledge and pedagogical content knowledge needs in a physics activity. *Teaching and Teacher Education, 22*, 503–512; Menon, D., & Sadler, T. D. (2016). Preservice elementary teachers' science self-efficacy beliefs and science content knowledge. *Journal of Science Teacher Education, 27*, 649–673.

6. Donaldson, M. (1979). The mismatch between school and children's minds. *Human Nature, 2*, 60–67; Tzard, B., & Hughes, M. (1984). *Young children learning*. Harvard University Press.

7. Beeth, M. E., & Hewson, P. W. (1999). Learning goals in an exemplary science teacher's practice: Cognitive and social factors in teaching for conceptual change. *Science Education, 83*, 738–760,

8. See also: Smith, C. L., Maclin, D., Houghton, C., & Hennessey, M. G. (2000). Sixth-grade students' epistemologies of science: The impact of school science experiences on epistemological development. *Cognition and Instruction, 18*, 349–422.

9. Hennessey, M. G. (2008). A case study of a student's reflective thoughts: A vision for practice. In The Role of Metacognition in Teaching Geoscience Workshop, November 19-21, 2008, at Carleton College, Northfield, MN. National Association of Geoscience Teachers. https://serc.carleton.edu/NAGTWorkshops/metacognition/workshop08/participants/hennessey.html.

10. National Research Council. (2007). *Taking science to school: Learning and teaching science in grades K–8*. National Academies Press.

11. See Klahr, D., Zimmerman, C., & Jirout, J. (2011). Educational interventions to advance children's scientific thinking. *Science, 333*(6045), 971–975.

12. Conant, J. B. (1947). *On understanding science: An historical approach*. Yale University Press; Conant, J. B., & Nash, L. K. (Eds.). (1957). *Harvard case histories in experimental science*. Harvard University Press.

13. Eichman, P. (1996). Using history to teach biology. *The American Biology Teacher, 58*(4), 200–204; Wang, H. A., & Marsh, D. D. (2002). Science instruction with a humanistic twist: Teachers' perception and practice in using the history of science in their classrooms. *Science & Education, 11*(2), 169–189; Monk, M., & Osborne, J. (1997). Placing the history and philosophy of science on the curriculum: A model for the development of pedagogy. *Science Education, 81*, 405–424; Lin, C. Y., Cheng, J. H., & Chang, W. H. (2010). Making science vivid: Using a historical episodes map. *International Journal of Science Education, 32*, 2521–2531.

14. Hong, H.-Y., & Lin-Siegler, X. (2012). How learning about scientists' struggles influences students' interest and learning in physics. *Journal of Educational Psychology, 104*, 469–484; Allchin, D., Andersen, H. M., & Nielsen, K. (2014). Complementary approaches to teaching nature of science: Integrating student inquiry, contemporary cases and historical cases in classroom practice. *Science Education, 98*, 461–486.

15. DeCaro, M. S., & Rittle-Johnson, B. (2012). Exploring mathematics problems prepares children to learn from instruction. *Journal of Experimental Child Psychology, 113*, 552–568.

16. Monk, M., & Osborne, J. (1997). Placing the history and philosophy of science on the curriculum: A model for the development of pedagogy. *Science Education, 81*, 405–424.

17. E.g., Wang, H. A., & Marsh, D. D. (2002). Science instruction with a humanistic twist: teachers' perception and practice in using the history of science in their classrooms. *Science & Education, 11*(2), 169–189; Jenkins, E. W. (2013). The "nature of

science" in the school curriculum: The great survivor. *Journal of Curriculum Studies, 45*, 132–151.

18. Klahr et al. (2011), Educational interventions to advance children's scientific thinking.

19. Klahr et al. (2011), Educational interventions to advance children's scientific thinking.

20. Organisation for Economic Co-operation and Development (OECD). (2007). *PISA 2006: Science competencies for tomorrow's world* (Vol. 1); OECD (2008). *Education at a glance: OECD indicators 2008*; Sahlberg, P. (2015). *Finnish lessons 2.0: What can the world learn from educational change in Finland?* Teachers College Press; Sahlberg, P. (2016). The global educational reform movement and its impact on schooling. In K. E. Mundy, A. Green, B. Lingard, & A. Verger (Eds.), *The handbook of global education policy* (pp. 128–144). Wiley; Sahlberg, P. (2021). *Finnish lessons 3.0: What can the world learn from educational change in Finland?* Teachers College Press.

21. PISA scores dropped somewhat in recent years, possibly because of the influence of screen time on children and young teens and because some countries are teaching to the PISA test. But, in broader comparisons, Finland remains in the top ranks, Sahlberg, P., & Doyle, W. (2019). *Let the children play: How more play will save our schools and help children thrive.* Oxford University Press.

22. Ripley, A. (2013). *The smartest kids in the world: And how they got that way.* Simon & Schuster; Hancock, L. (2011, September). Why are Finland's schools successful? *Smithsonian Magazine*; Woessmann, L. (2016). The importance of school systems: Evidence from international differences in student achievement. *Journal of Economic Perspectives, 30*(3), 3–32.

23. Many other reports echo Ripley's observations; even as they vary in stressing what matters, they all tend to agree that the transformation of teaching into an elite profession was central, OECD (2007). *No more failures: Ten steps to equity in education*; Maaranen, K., Kynäslahti, H., Byman, R., Jyrhämä, R., & Sintonen, S. (2019). Teacher education matters: Finnish teacher educators' concerns, beliefs, and values. *European Journal of Teacher Education, 42*(2), 211–227; Mikkilä-Erdmann, M., Warinowski, A., & Iiskala, T. (2019, May 23). Teacher education in Finland and future directions. *Oxford research encyclopedia of education.* https://oxfordre.com/education/view/10.1093/acrefore/9780190264093.001.0001/acrefore-9780190264093-e-286.

24. Darling-Hammond, L. (2010). *The flat world and education: How America's commitment to equity will determine our future.* Teachers College Press.

25. Hancock (2011), Why are Finland's schools successful?

26. Sahlberg (2015), *Finnish lessons 2.0.*

27. Ripley (2013). *The smartest kids in the world*; Sahlberg & Doyle (2019), *Let the children play.*

28. Will, M. (2018, August 9). Enrollment is down at teacher colleges: So they're trying to change. *Education Week.*

29. Sahlberg (2015), *Finnish lessons 2.0.*

30. Saavedra, J., Alasuutari, H, & Gutierrez, M. (2018, August 14). Finland's education system: The journey to success. World Bank blog.

31. Sahlberg & Doyle (2019), *Let the children play.*

32. Chu, J., & Schulz, L. E. (2020). Play, curiosity, and cognition. *Annual Review of Developmental Psychology, 2,* 317–343; Bruner. J. S., Jolly, A., & Sylva, K., eds. (1976). *Play: Its role in development and evolution.* Penguin; Lillard, A. S. (2015). The development of play. In L. S. Liben, U. Müller& R. M. Lerner (Eds.), *Handbook of child psychology and developmental science, vol. 2: Cognitive processes* (7th ed.) (pp. 425–468). John Wiley; Martin, P., & Caro, T. M. (1985). On the functions of play and its role in behavioral development. *Advances in the Study of Behavior, 15,* 59–103.

33. Sahlberg & Doyle (2019), *Let the children play.*

34. Riede, F., Johannsen, N. N., Högberg, A., Nowell, A., & Lombard, M. (2018). The role of play objects and object play in human cognitive evolution and innovation. *Evolutionary Anthropology: Issues, News, and Reviews, 27*(1), 46–59.

35. For example, Ashcraft, M. H. (2002). Math anxiety: Personal, educational, and cognitive consequences. *Current Directions in Psychological Science, 11,* 181–185; Ramirez, G., Shaw, S. T., & Maloney, E. A. (2018). Math anxiety: Past research, promising interventions, and a new interpretation framework. *Educational Psychologist, 53,* 145–164.

36. Yousif, S. R., & Keil, F. C. (2020). Area, not number, dominates estimates of visual quantities. *Scientific Reports, 10,* 1–13.

37. Ramirez, G., Chang, H., Maloney, E. A., Levine, S. C., & Beilock, S. L. (2016). On the relationship between math anxiety and math achievement in early elementary school: The role of problem solving strategies. *Journal of Experimental Child Psychology, 141,* 83–100.

38. See also Chang, H., & Beilock, S. L. (2016). The math anxiety-math performance link and its relation to individual and environmental factors: A review of current behavioral and psychophysiological research. *Current Opinion in Behavioral Sciences, 10,* 33–38.

39. Schoenfeld, A. H. (2013). *Cognitive science and mathematics education.* Routledge.

40. Hewitt, P. G. (2009). *Conceptual physics: The high school physics program* (3rd ed.). Pearson/Prentice Hall; Shapiro, G. (1979). *Physics without math: A descriptive introduction.* Prentice Hall.

41. Boaler, J. (2019). Developing mathematical mindsets: The need to interact with numbers flexibly and conceptually. *American Educator, 42,* 28–33.

42. Fuchs, L. S., Gilbert, J. K., Powell, S. R., Cirino, P. T., Fuchs, D., Hamlett, C. L., . . . Tolar, T. D. (2016). The role of cognitive processes, foundational math skill, and calculation accuracy and fluency in word-problem solving versus prealgebraic knowledge. *Developmental Psychology, 52*, 2085–2098; Mädamürk, K., Kikas, E., & Palu, A. (2018). Calculation and word problem-solving skill profiles: Relationship to previous skills and interest. *Educational Psychology, 38*, 1239–1254.

43. Foley, A. E., Herts, J. B., Borgonovi, F., Guerriero, S., Levine, S. C., & Beilock, S. L. (2017). The math anxiety-performance link: A global phenomenon. *Current Directions in Psychological Science, 26*, 52–58.

44. Sorvo, R., Koponen, T., Viholainen, H., Aro, T., Räikkönen, E., Peura, P., . . . Aro, M. (2017). Math anxiety and its relationship with basic arithmetic skills among primary school children. *British Journal of Educational Psychology, 87*, 309–327.

45. Wang, M. T., & Degol, J. L. (2017). Gender gap in science, technology, engineering, and mathematics (STEM): Current knowledge, implications for practice, policy, and future directions. *Educational Psychology Review, 29*, 119–140; Hyde, J. S., & Linn, M. C. (2006). Gender similarities in mathematics and science. *Science, 314*(5799), 599–600.

46. Lei, R. F., Green, E. R., Leslie, S. J., & Rhodes, M. (2019). Children lose confidence in their potential to "be scientists," but not in their capacity to "do science." *Developmental Science, 22*, e12837.

47. Lepper, M. R., Greene, D., & Nisbett, R. E. (1973). Undermining children's intrinsic interest with external reward: A test of the undermining hypothesis. *Journal of Personality and Social Psychology, 28*, 129–147.

48. For example, Lepper, M. R., Corpus, J. H., & Iyengar, S. S. (2005). Intrinsic and extrinsic motivational orientations in the classroom: Age differences and academic correlates. *Journal of Educational Psychology, 97*, 184–196; Deci, E. L., Koestner, R., & Ryan, R. M. (1999). A meta-analytic review of experiments examining the effects of extrinsic rewards on intrinsic motivation. *Psychological Bulletin, 125*, 627–668.

49. Amabile, T. M. (1996). *Creativity in context*. Westview Press; Byron, K., & Khazanchi, S. (2012). Rewards and creative performance: A meta-analytic test of theoretically derived hypotheses. *Psychological Bulletin, 138*, 809–830.

50. Lepper, M. R., Corpus, J. H., & Iyengar, S. S. (2005). Intrinsic and extrinsic motivational orientations in the classroom: Age differences and academic correlates. *Journal of Educational Psychology, 97*, 184–196.

51. Kohn, A. (1993). *Punished by rewards: The trouble with gold stars, incentive plans, A's, praise, and other bribes*. Houghton Mifflin.

52. Dweck, C. S. (2008). *Mindset: The new psychology of success*. Random House Digital.

53. Cimpian, A., Arce, H.-M. C., Markman, E. M., & Dweck, C. S. (2007). Subtle linguistic cues affect children's motivation. *Psychological Science, 18*, 314–316; Corpus, J. H., & Lepper, M. R. (2007). The effects of person versus performance praise on children's motivation: Gender and age as moderating factors. *Educational Psychology, 27*, 487–508; Zentall, S. R., & Morris, B. J. (2010). "Good job, you're so smart": The effects of inconsistency of praise type on young children's motivation. *Journal of Experimental Child Psychology, 107*, 155–163.

54. Gunderson, E. A., Gripshover, S. J., Romero, C., Dweck, C. S., Goldin-Meadow, S., & Levine, S. C. (2013). Parent praise to 1- to 3-year-olds predicts children's motivational frameworks 5 years later. *Child Development, 84*, 1526–1541.

55. Yeager, D. S., Hanselman, P., Walton, G. M., Murray, J. S., Crosnoe, R., Muller, C., . . . Paunesku, D. (2019). A national experiment reveals where a growth mindset improves achievement. *Nature, 573*(7774), 364–369; Yeager, D. S., & Walton, G. M. (2011). Social-psychological interventions in education: They're not magic. *Review of Educational Research, 81*, 267–301.

56. Sisk, V. F., Burgoyne, A. P., Sun, J., Butler, J. L., & Macnamara, B. N. (2018). To what extent and under which circumstances are growth mind-sets important to academic achievement? Two meta-analyses. *Psychological Science, 29*(4), 549–571; Burgoyne, A. P., Hambrick, D. Z., & Macnamara, B. N. (2020). How firm are the foundations of mind-set theory? The claims appear stronger than the evidence. *Psychological Science, 31*, 258–267.

57. Lockhart, K. L., Chang, B., & Story, T. (2002). Young children's beliefs about the stability of traits: Protective optimism? *Child Development, 73*, 1408–1430; Lockhart et al. (2008), From ugly duckling to swan?

58. See also Boseovski, J. J. (2010). Evidence for "rose-colored glasses": An examination of the positivity bias in young children's personality judgments. *Child Development Perspectives, 4*, 212–218; Diesendruck, G., & Lindenbaum, T. (2009). Self-protective optimism: Children's biased beliefs about the stability of traits. *Social Development, 18*, 946–961.

59. Lockhart et al. (2002), Young children's beliefs about the stability of traits; see also Cole, D. A., Ciesla, J. A., Dallaire, D. H., Jacquez, F. M., Pineda, A. Q., LaGrange, B., . . . Felton, J. W. (2008). Emergence of attributional style and its relation to depressive symptoms. *Journal of Abnormal Psychology, 117*, 16–31.

60. Lockhart, K. L., Goddu, M. K., & Keil, F. C. (2017). Overoptimism about future knowledge: Early arrogance? *The Journal of Positive Psychology, 12*, 36–46.

61. Kidd, C., & Hayden, B. Y. (2015). The psychology and neuroscience of curiosity. *Neuron, 88*, 449–460; Jirout, J., & Klahr, D. (2012). Children's scientific curiosity: In search of an operational definition of an elusive concept. *Developmental Review,*

32, 125–160; Piotrowski, J. T., Litman, J. A., & Valkenburg, P. (2014). Measuring epistemic curiosity in young children. *Infant and Child Development, 23*, 542–553.

62. Berlyne, D. E. (1954). A theory of human curiosity. *British Journal of Psychology, 45*, 180–191.

63. Engel, S. (2011). Children's need to know: Curiosity in schools. *Harvard Educational Review, 81*(4), 625–645.

64. Klahr et al. (2011), Educational interventions to advance children's scientific thinking; Bathgate et al. (2014), Children's motivation toward science across contexts, manner of interaction, and topic.

65. National Research Council. (2012). *A framework for K–12 science education: Practices, crosscutting concepts, and core ideas.* National Academies Press.

66. Bathgate, M., Crowell, A., Schunn, C., Cannady, M., & Dorph, R. (2015). The learning benefits of being willing and able to engage in scientific argumentation. *International Journal of Science Education, 37*, 1590–1612; Osborne, J. F., Borko, H., Fishman, E., Gomez Zaccarelli, F., Berson, E., Busch, K. C., . . . Tseng, A. (2019). Impacts of a practice-based professional development program on elementary teachers' facilitation of and student engagement with scientific argumentation. *American Educational Research Journal, 56*, 1067–1112.

67. Dewey, J. (2010). To those who aspire to the profession of teaching (APT). In D. J. Simpson & S. F. Stack Jr. (Eds.), *Teachers, leaders and schools: Essays by John Dewey* (pp. 33–36). Southern Illinois University Press.

Chapter 7

1. Coley, J. D., & Tanner, K. (2015). Relations between intuitive biological thinking and biological misconceptions in biology majors and nonmajors. *CBE—Life Sciences. Education, 14*, 8; Blancke, S., Tanghe, K. B., & Braeckman, J. (2018). Intuitions in science education and the public understanding of science. In K. Rutten, S. Blancke, & R. Soetaert (Eds.), *Perspectives on science and culture* (pp. 223–241). Purdue University Press.

2. Gelman, S. A. (2003). *The essential child: Origins of essentialism in everyday thought.* Oxford University Press; Barrett, H. C. (2001). On the functional origins of essentialism. *Mind & Society, 2*, 1–30; F. C. Keil (1989). *Concepts, Kinds and Cognitive Development.* MIT Press.

3. Hull, D. L. (1965). The effect of essentialism on taxonomy: 2000 years of stasis. *British Journal of Philosophical Science, 15*, 314–326; Gelman, S. A., & Rhodes, M. (2012). Two-thousand years of stasis. In K. S. Rosengren, S. K. Brem, E. M. Evans, & G. M. Sinatra (Eds.), *Evolution challenges: Integrating research and practice in teaching and learning about evolution* (pp. 200–207). Oxford University Press.

4. Many scholars see essentialism as wrong even for natural kinds and for causing false impressions of biological differences among races and gender, for example, Leslie, S. J. (2013). Essence and natural kinds: When science meets preschooler intuition. *Oxford Studies in Epistemology, 4*, 108–165.

5. Noyes, A., & Keil, F. C. (2017). Revising deference: Intuitive beliefs about category structure constrain expert deference. *Journal of Memory and Language, 95*, 68–77.

6. Noyes, A., & Keil, F. C. (2019). Generics designate kinds but not always essences. *Proceedings of the National Academy of Sciences, 116*, 20354–20359.

7. Noyes, A., Keil, F. C. (2020). There is no privileged link between kinds and essences early in development. *Proceedings of the National Academy of Sciences, 117*, 10633–10635.

8. Shtulman, A. (2017). *Scienceblind: Why our intuitive theories about the world are so often wrong*. Basic Books; Shtulman, A. (2019). Doubly counterintuitive: Cognitive obstacles to the discovery and the learning of scientific ideas and why they often differ. In R. Samuels & D. Wilkenfeld (Eds.), *Advances in experimental philosophy of science* (pp. 97–121). Bloomsbury.

9. Shtulman, A., & Valcarcel, J. (2012). Scientific knowledge suppresses but does not supplant earlier intuitions. *Cognition, 124*, 209–215; Wandersee, J. H. (1986). Can the history of science help science educators anticipate students' misconceptions? *Journal of Research in Science Teaching, 23*, 581–597.

10. Sinatra, G. M., Kienhues, D., & Hofer, B. K. (2014). Addressing challenges to public understanding of science: Epistemic cognition, motivated reasoning, and conceptual change. *Educational Psychologist, 49*, 123–138; Gervais, W. M. (2015). Override the controversy: Analytic thinking predicts endorsement of evolution. *Cognition, 142*, 312–321.

11. Schwartz, D. L., Tsang, J. M., & Blair, K. P. (2016). *The ABCs of how we learn: 26 scientifically proven approaches, how they work, and when to use them*. W. W. Norton.

12. Shtulman (2017), *Scienceblind*; Potvin, P., Masson, S., Lafortune, S., & Cyr, G. (2015). Persistence of the intuitive conception that heavier objects sink more: A reaction time study with different levels of interference. *International Journal of Science and Mathematics Education, 13*(1), 21–43; Goldberg, R. F., & Thompson-Schill, S. L. (2009). Developmental "roots" in mature biological knowledge. *Psychological Science, 20*(4), 480–487.

13. Gregory, T. R. (2009). Artificial selection and domestication: Modern lessons from Darwin's enduring analogy. *Evolution: Education and Outreach, 2*, 5–27.

14. Ordás, A., & Cartea, M. E. (2008). Cabbage and kale. In J. Prohens (Ed.), *Vegetables I* (pp. 119–149). Springer; image at https://evolution.berkeley.edu/evolibrary/article/evo_30.

15. Bull, J. W., & Maron, M. (2016). How humans drive speciation as well as extinction. *Proceedings of the Royal Society B: Biological Sciences, 283,* 20160600.

16. Kelemen, D., Emmons, N. A., Seston Schillaci, R., & Ganea, P. A. (2014). Young children can be taught basic natural selection using a picture-storybook intervention. *Psychological Science, 25,* 893–902; Emmons, N., Lees, K., & Kelemen, D. (2018). Young children's near and far transfer of the basic theory of natural selection: An analogical storybook intervention. *Journal of Research in Science Teaching, 55,* 321–347.

17. Kelemen, D. (2019). The magic of mechanism: Explanation-based instruction on counterintuitive concepts in early childhood. *Perspectives on Psychological Science, 14,* 510–522.

18. Jacoby, L. L., Kelley, C., Brown, J., & Jasechko, J. (1989). Becoming famous overnight: Limits on the ability to avoid unconscious influences of the past. *Journal of Personality and Social Psychology, 56,* 326–338; see Schacter, D. L. (Ed.). (1997). *Memory distortion: How minds, brains, and societies reconstruct the past.* Harvard University Press.

19. Shtulman (2019), Doubly counterintuitive.

20. Blancke, S., & De Smedt, J. (2013). Evolved to be irrational? Evolutionary and cognitive foundations. In M. Pigliucci & M. Boudry (Eds.), *The philosophy of pseudoscience. Reconsidering the demarcation problem* (pp. 361–379). University of Chicago Press; Stanovich, K. E. (2018). Miserliness in human cognition: The interaction of detection, override and mindware. *Thinking & Reasoning, 24,* 423–444.

21. Kelemen (2019), The magic of mechanism.

22. Stanovich (2018), Miserliness in human cognition.

23. Legare, C. H., Evans, E. M., Rosengren, K. S., & Harris, P. L. (2012). The coexistence of natural and supernatural explanations across cultures and development. *Child Development, 83,* 779–793; Shtulman, A., & Legare, C. H. (2020). Competing explanations of competing explanations: Accounting for conflict between scientific and folk explanations. *Topics in Cognitive Science, 12*(4), 1337–1362.

24. For example, Collins, F. S. (2006). *The language of God: A scientist presents evidence for belief.* Simon & Schuster; versus Dawkins, R. (2006). *The God delusion.* Houghton Mifflin Co.

25. For an illustration of related effects see: Johnson, S. G. B., Zhang, J., & Keil, F. C. (2021). Win–win denial: The psychological underpinnings of zero-sum thinking. *Journal of Experimental Psychology: General.* https://doi.org/10.1037/xge0001083.

26. Shtulman, A., & Legare, C. H. (2020). Competing explanations of competing explanations: accounting for conflict between scientific and folk explanations. *Topics in Cognitive Science, 12,* 1337–1362.

27. Kahneman, D. (2011). *Thinking, fast and slow*. Macmillan; Sloman, S. A. (2002). Two systems of reasoning. In T. Gilovich, D. Griffin, & D. Kahneman (Eds.), *Heuristics and biases: The psychology of intuitive judgment* (pp. 379–396). Cambridge University Press; Blancke & De Smedt (2013), Evolved to be irrational?

28. Stanovich (2018), Miserliness in human cognition.

29. Stanovich (2018), Miserliness in human cognition.

30. Risen, J. L. (2016). Believing what we do not believe: Acquiescence to superstitious beliefs and other powerful intuitions. *Psychological Review, 123*, 182–207.

31. Rozenblit, L., & Keil, F. C. (2002). The misunderstood limits of folk science: An illusion of explanatory depth. *Cognitive Science, 26*, 521–562.

32. Rozenblit, L., & Keil, F. C. (2002). The misunderstood limits of folk science.

33. Alter, A. L., Oppenheimer, D. M., & Zemla, J. C. (2010). Missing the trees for the forest: A construal level account of the illusion of explanatory depth. *Journal of Personality and Social Psychology, 99*, 436–451; Fernbach, P. M., Rogers, T., Fox, C. R., & Sloman, S. A. (2013). Political extremism is supported by an illusion of understanding. *Psychological Science, 24*, 939–946; Fisher, M., Goddu, M. K., & Keil, F. C. (2015). Searching for explanations: How the internet inflates estimates of internal knowledge. *Journal of Experimental Psychology: General, 144*, 674–687; Vitriol, J. A., & Marsh, J. K. (2018). The illusion of explanatory depth and endorsement of conspiracy beliefs. *European Journal of Social Psychology, 48*, 955–969.

34. Dunning, D. (2011). The Dunning–Kruger effect: On being ignorant of one's own ignorance. In *Advances in social psychology* (Vol. 44, pp. 247–296). Academic Press.

35. Sloman, S., & Fernbach, P. (2018). *The knowledge illusion: Why we never think alone*. Penguin; Sloman, S. A., & Rabb, N. (2016). Your understanding is my understanding: Evidence for a community of knowledge. *Psychological Science, 27*, 1451–1460.

36. Clark, A., & Chalmers, D. (1998). The extended mind. *Analysis, 58*, 7–19.

37. Wegner, D. M. (1987). Transactive memory: A contemporary analysis of the group mind. In B. Mullen & G. R. Goethals (Eds.), *Theories of group behavior* (pp. 185–208). Springer; Harris, C., Barnier, A., Sutton, J., & Keil, P. (2014). Couples as socially distributed cognitive systems: Remembering in everyday social and material contexts. *Memory Studies, 7*, 285–297.

38. Mills, C., & Keil, F. C. (2004). Knowing the limits of one's understanding: The development of an awareness of an illusion of explanatory depth. *Journal of Experimental Child Psychology, 87*, 1–32.

39. Wilson, R. A., & Keil, F. (1998). The shadows and shallows of explanation. *Minds and Machines, 8*, 137–159.

40. Lockhart, K. L., Chang, B., & Story, T. (2002). Young children's beliefs about the stability of traits: Protective optimism?. *Child Development, 73,* 1408–1430; see also Bjorklund, D. F. (1997). The role of immaturity in human development. *Psychological Bulletin, 122,* 153–169.

41. Sloman & Fernbach (2018), *The knowledge illusion.*

42. Dunning (2011), The Dunning–Kruger effect; Hamilton, E., Cairns, H., & Cooper, L. (1961). *The collected dialogues of Plato.* Princeton University Press.

43. Weisberg, D. S., Keil, F. C., Goodstein, J., Rawson, E., & Gray, J. R. (2008). The seductive allure of neuroscience explanations. *Journal of Cognitive Neuroscience, 20,* 470–477.

44. Keil, F. C., Lockhart, K. L., & Schlegel, E. (2010). A bump on a bump? Emerging intuitions concerning the relative difficulty of the sciences. *Journal of Experimental Psychology: General, 139,* 1–15.

45. Fodor, J. A. (1975). *The language of thought.* Harvard University Press.

46. Fisher et al. (2015), Searching for explanations.

47. Isaacson, W. (2017). *Leonardo da Vinci.* Simon & Schuster (Kindle ed.).

48. Dolnick, E. (2011). *The Clockwork Universe: Isaac Newton, the Royal Society, and the Birth of the Modern World I.* HarperCollins.

49. Robinson, A. (2007). *The last man who knew everything.* Pi Press.

50. Robinson (2007), *The last man who knew everything,* p. 14.

51. Robinson (2007), *The last man who knew everything,* p. 19.

52. See Richard Feynman (n.d). Wikiquotes. https://en.wikiquote.org/wiki/Richard_Feynman.

53. See related argument in Root-Bernstein, R. (2009). Multiple giftedness in adults: The case of polymaths. In L. V. Shavinina (Ed.), *International handbook on giftedness* (pp. 853–870). Springer.

54. January 30, 2018, Smriti Mallapaty in *Nature* Index.

55. Aboukhalil, R. (2014). The rising trend in authorship. *The Winnower, 8,* e141832.26907; see also: Chawla, D. S. (2019, December 13). Hyperauthorship: global projects spark surge in thousand-author papers. *Nature.* https://www.nature.com/articles/d41586-019-03862-0; Nielsen, M. W., & Andersen, J. P. (2021). Global citation inequality is on the rise. *Proceedings of the National Academy of Sciences, 118,* e2012208118.

Chapter 8

1. Hossenfelder, S. (2019, June 12). Is climate change inconvenient or existential? Only supercomputers can do the math. *New York Times*, includes interview with the Oxford geophysicist Tim Palmer.

2. Ranney, M. A., & Clark, D. (2016). Climate change conceptual change: Scientific information can transform attitudes. *Topics in Cognitive Science, 8*, 49–75.

3. Shtulman, A. (2017). *Scienceblind: Why our intuitive theories about the world are so often wrong*. Basic Books.

4. Cook, J., Nuccitelli, D., Green, S. A., Richardson, M., Winkler, B., Painting, R., . . . & Skuce, A. (2013). Quantifying the consensus on anthropogenic global warming in the scientific literature. *Environmental Research Letters, 8*, 024024.

5. Palmer, T., & Stevens, B. (2019). The scientific challenge of understanding and estimating climate change. *Proceedings of the National Academy of Sciences, 116*, 24390–24395.

6. Kahan, D. (2017). On the sources of ordinary science knowledge and extraordinary science ignorance. In. K. H. Jamieson, D. M. Kahan, & D. Scheufele (Eds.). In *The Oxford handbook of the science of science communication* (pp. 35–49). Oxford University Press.

7. Ranney & Clark (2016), Climate change conceptual change.

8. Fernbach, Rogers, Fox, & Sloman (2013), Political extremism is supported by an illusion of understanding; Fernbach, P. M., Sloman, S. A., Louis, R. S., & Shube, J. N. (2013). Explanation fiends and foes: How mechanistic detail determines understanding and preference. *Journal of Consumer Research, 39*, 1115-1131.

9. Van der Linden, S., Leiserowitz, A., Rosenthal, S., & Maibach, E. (2017). Inoculating the public against misinformation about climate change. *Global Challenges, 1*, 1600008.

10. Fernbach, P. M., Light, N., Scott, S. E., Inbar, Y., & Rozin, P. (2019). Extreme opponents of genetically modified foods know the least but think they know the most. *Nature Human Behaviour, 3*, 251–256.

11. Fernbach et al. (2019), Extreme opponents of genetically modified foods know the least but think they know the most.

12. Shtulman, A. (2017). *Scienceblind*; Kelemen, D. (2019). The magic of mechanism: Explanation-based instruction on counterintuitive concepts in early childhood. *Perspectives on Psychological Science, 14*, 510–522.

13. Tarazona, J. V., Court-Marques, D., Tiramani, M., Reich, H., Pfeil, R., Istace F., & Crivellente F. (2017). Glyphosate toxicity and carcinogenicity: A review of the scientific basis of the European Union assessment and its differences with IARC. *Archives*

of Toxicology, 91, 2723–2743; *Wall Street Journal*. (2020, June 26). The Roundup settlement.

14. Steel, S. (2015). *Proof of causation in tort law*. Cambridge University Press; Gold, S. C. (2013). When certainty dissolves into probability: A legal vision of toxic causation for the post-genomic era. *Washington and Lee Law Review, 70*, 237–339.

15. Grover, P. L. (Ed.). (2019). *Chemical carcinogens & DNA* (Vol. 2). CRC Press; Kobets, T., Iatropoulos, M. J., & Williams, G. M. (2019). Mechanisms of DNA-reactive and epigenetic chemical carcinogens: Applications to carcinogenicity testing and risk assessment. *Toxicology Research, 8*, 123–145.

16. Birkett, N., Al-Zoughool, M., Bird, M., Baan, R. A., Zielinski, J., & Krewski, D. (2019). Overview of biological mechanisms of human carcinogens. *Journal of Toxicology and Environmental Health, Part B, 22*, 288–359.

17. Weisberg, D. S., Landrum, A. R., Metz, S. E., & Weisberg, M. (2018). No missing link: Knowledge predicts acceptance of evolution in the United States. *Bioscience, 68*, 212–222; Legare, C. H., Opfer, J. E., Busch, J. T., & Shtulman, A. (2018). A field guide for teaching evolution in the social sciences. *Evolution and Human Behavior, 39*, 257–268.

18. See National Institutes of Health factsheet listing of daily milligrams of vitamin C needed by various groups. https://ods.od.nih.gov/factsheets/VitaminC-HealthProfessional.

19. Kantor, E. D., Rehm, C. D., Du, M., White, E., & Giovannucci, E. L. (2016). Trends in dietary supplement use among US adults from 1999–2012. *JAMA, 316*, 1464–1474. https://www.statista.com/statistics/235801/retail-sales-of-vitamins-and-nutritional-supplements-in-the-us.

20. Cowan, A. E., Jun, S., Gahche, J. J., Tooze, J. A., Dwyer, J. T., Eicher-Miller, H. A., . . . Bailey, R. L. (2018). Dietary supplement use differs by socioeconomic and health-related characteristics among US adults, NHANES 2011–2014. *Nutrients, 10*, 1114; Ronis, M. J., Pedersen, K. B., & Watt, J. (2018). Adverse effects of nutraceuticals and dietary supplements. *Annual Review of Pharmacology and Toxicology, 58*, 583–601.

21. Teichner, W. & Lesko. M. (2013, December 1). Cashing in on the booming market for dietary supplements. *McKinsey Insights*. https://www.mckinsey.com/business-functions/marketing-and-sales/our-insights/cashing-in-on-the-booming-market-for-dietary-supplements#; Ng, S., & Rockoff, J. (2013, March 31). With top lines drooping, firms reach for vitamins. *Wall Street Journal*, 1–4.

22. Shahbandeh,M. (2019, August 9). U.S. sales of vitamins and nutritional supplements manufacturing 2018–2019. Statista. https://www.statista.com/statistics/235801/retail-sales-of-vitamins-and-nutritional-supplements-in-the-us: Guallar, E., Stranges, S., Mulrow, C., Appel, L. J., & Miller, E. R. (2013). Enough is enough: Stop wasting money on vitamin and mineral supplements. *Annals of Internal Medicine, 159*(12),

850–851; Price, C. (2016). *Vitamania: How vitamins revolutionized the way we think about food*. Penguin Books; Swanson, E. S. (2016). *Science and society* (see chapter 3, Pseudoscience, pp. 55–84). Springer.

23. Hoover, N., Aguiniga, A., & Hornecker, J. (2019). In the adult population, does daily multivitamin intake reduce the risk of mortality compared with those who do not take daily multivitamins? *Evidence-Based Practice, 22*, 15.

24. Eisenberg, M. D., Avery, R. J., & Cantor, J. H. (2017). Vitamin panacea: Is advertising fueling demand for products with uncertain scientific benefit? *Journal of Health Economics, 55*, 30–44.

25. Ronis et al. (2018), Adverse effects of nutraceuticals and dietary supplements.

26. Pauling, L. (1971). Vitamin C and common cold. *JAMA, 216*, 332–332; Offit, P. (2013, July 19). The vitamin myth: Why we think we need supplements. *The Atlantic*.

27. Cowan et al. (2018), Dietary supplement use differs by socioeconomic and health-related characteristics among US adults, NHANES 2011–2014.

28. Lockhart, K. L., & Keil, F. C. (2019). What heals and why? *Monographs of the Society for Research in Child Development, 83*(2).

29. Kunda, Z. (1990). The case for motivated reasoning. *Psychological Bulletin, 108*, 480–498; Hart, P. S., & Nisbet, E. C. (2012). Boomerang effects in science communication: How motivated reasoning and identity cues amplify opinion polarization about climate mitigation policies. *Communication Research, 39*, 701–723.

30. Hart & Nisbet (2012). Boomerang effects in science communication; but see Druckman, J. N., & McGrath, M. C. (2019). The evidence for motivated reasoning in climate change preference formation. *Nature Climate Change, 9*, 111–119.

31. Druckman, J. N., & Bolsen, T. (2011). Framing, motivated reasoning, and opinions about emergent technologies. *Journal of Communication, 61*, 659–688.

32. Helzer, E. G., & Dunning, D. (2012). On motivated reasoning and self-belief. In S. Vazire & T. D. Wilson (Eds.), *Handbook of self-knowledge* (pp. 379–396). Guilford Press.

33. Bernard, S., Clément, F., & Mercier, H. (2016). Wishful thinking in preschoolers. *Journal of Experimental Child Psychology, 141*, 267–274.

34. Mills, C. M., & Keil, F. C. (2005). The development of cynicism. *Psychological Science, 16*, 385–390.

35. Lockhart, K. L., Goddu, M. K., & Keil, F. C. (2018). When saying "I'm best" is benign: Developmental shifts in perceptions of boasting. *Developmental Psychology, 54*, 521–535.

36. Asch, S. E. (1956). Studies of independence and conformity: I. A minority of one against a unanimous majority. *Psychological Monographs: General and Applied, 70*, 1–70.

37. Kahan, D. M., Jenkins-Smith, H., & Braman, D. (2011). Cultural cognition of scientific consensus. *Journal of Risk Research, 14*, 147–174.

38. Esposo, S. R., Hornsey, M. J., & Spoor, J. R. (2013). Shooting the messenger: Outsiders critical of your group are rejected regardless of argument quality. *British Journal of Social Psychology, 52*, 386–395; Kahan, D. M., Braman, D., Cohen, G. L., Gastil, J., & Slovic, P. (2010). Who fears the HPV vaccine, who doesn't, and why? An experimental study of the mechanisms of cultural cognition. *Law and Human Behavior, 34*, 501–516.

39. Kahan, D. M. (2017). 'Ordinary science intelligence': A science-comprehension measure for study of risk and science communication, with notes on evolution and climate change. *Journal of Risk Research, 20*, 995–1016.

40. McPhetres, J., & Pennycook, G. (2019). Science beliefs, political ideology, and cognitive sophistication. https://osf.io/ad9v7.

41. Lewandowsky, S. (2021). Climate change disinformation and how to combat it. *Annual Review of Public Health, 42*, 1–21.

42. Stanchevici, D. (2017). *Stalinist genetics: The constitutional rhetoric of TD Lysenko*. Routledge.

43. Kean, S. (2017, December 19). The Soviet era's deadliest scientist is regaining popularity in Russia. *The Atlantic*; Stanchevici (2017), *Stalinist genetics*.

44. Becker, J. (1998). *Hungry ghosts: Mao's secret famine*. Macmillan; Zielinski, S. (2010, February 1). When the Soviet Union chose the wrong side on genetics and evolution. Smithsonianmag.com.

45. Reznik, S., & Fet, V. (2019). The destructive role of Trofim Lysenko in Russian science. *European Journal of Human Genetics, 27*, 1324–1325; Meloni, M. (2016). Rethinking Lysenko's legacy. *Science, 352*(6284), 421; Wang, Z., & Liu, Y. (2017). Lysenko and Russian genetics: An alternative view. *European Journal of Human Genetics, 25*, 1097–1098.

46. Corbyn, Z. (2010, June 22). AIDS researcher cleared of misconduct. *Nature*. https://doi.org/10.1038/news.2010.310.

47. Lewandowsky, S., & Oberauer, K. (2016). Motivated rejection of science. *Current Directions in Psychological Science, 25*, 217–222.

48. Lewandowsky & Oberauer (2016), Motivated rejection of science.

49. Kahan, D. M., Peters, E., Dawson, E. C., & Slovic, P. (2017). Motivated numeracy and enlightened self-government. *Behavioural public policy, 1*, 54-86; Nisbet, M. C. (2016). The science literacy paradox: Why really smart people often have the most biased opinions. *Skeptical Inquirer, 40*, 21–23.

50. McPhetres & Pennycook (2019), Science beliefs, political ideology, and cognitive sophistication.

51. Brashier, N. M., & Marsh, E. J. (2020). Judging truth. *Annual Review of Psychology, 71*, 499–515.

52. Pennycook, G., & Rand, D. G. (2019). Lazy, not biased: Susceptibility to partisan fake news is better explained by lack of reasoning than by motivated reasoning. *Cognition, 188*, 39–50.

53. Lewandowsky & Oberauer (2016), Motivated rejection of science.

54. Hart Research (2019, March 6). New poll: Overwhelming majority of Americans support more federal funding for science and technology research. https://hartresearch.com/?s=+funding+science; for more detailed report see: https://hartresearch.com/wp-content/uploads/2019/01/Federal-Science-Funding.pdf.

55. Interest and trust are not the same. The public's trust in science, while not great, has stayed roughly the same for several years.

56. For example, Tröbst, S., Kleickmann, T., Lange-Schubert, K., Rothkopf, A., & Möller, K. (2016). Instruction and students' declining interest in science: An analysis of German fourth- and sixth-grade classrooms. *American Educational Research Journal, 53*, 162–193; Kennedy, B., Hefferon, M., & Funk, C. (2018, January 17). Half of Americans think young people don't pursue STEM because it is too hard. Pew Research Center. https://www.pewresearch.org/fact-tank/2018/01/17/half-of-americans-think-young-people-dont-pursue-stem-because-it-is-too-hard.

57. Plumer, B. (2019, April 3). We fact-checked President Trump's dubious claims on the perils of wind power. *New York Times*. https://www.nytimes.com/2019/04/03/climate/fact-check-trump-windmills.html.

58. Reality Check team (2020, April 24). Coronavirus: Trump's disinfectant and sunlight claims fact-checked. BBC News. https://www.bbc.com/news/world-us-canada-52399464.

59. Kennedy, B., & Hefferon, M. (2019, March 28). What Americans know about science. Pew Research Center. https://www.pewresearch.org/science/2019/03/28/what-americans-know-about-science. See also Besley, J. C., & Hill, D. (2020, May 15). Science and technology: Public attitudes, knowledge, and interest. https://ncses.nsf.gov/pubs/nsb20207/executive-summary.

60. Brodeur, P. (1993). *The great power-line cover-up: How the utilities and the government are trying to hide the cancer hazard posed by electromagnetic fields*. Little, Brown.

61. Thomson, W. [Lord Kelvin] (2010). *Baltimore lectures on molecular dynamics and the wave theory of light*. Cambridge University Press. https://doi.org/10.1017/CBO9780511694523. (Originally published in 1904).

62. Thompson, S. P. (1910). *The life of William Thomson, Baron Kelvin of Largs* (Vol. 2), "Views and opinions." Macmillan.

Chapter 9

1. Thomas, A., & Chess, S. (1977). *Temperament and development.* Brunner/Mazel.

2. Falk, J. H., & Dierking, L. D. (2010). School is not where most Americans learn most of their science. *American Scientist, 98*, 486–493.

3. Khan, B., Robbins, C., Okren, A. (2020, January 15). The state of U.S. science and engineering 2020. National Science Foundation/National Science Board. https://ncses.nsf.gov/pubs/nsb20201.

4. Pinker, S. (2018). *Enlightenment now: The case for reason, science, humanism, and progress*. Penguin.

5. Ebeling, C. E. (2009). Evolution of a box. *Invention and Technology, 23*, 8–9.

6. Aman, M. M., Jasmon, G. B., Mokhlis, H., & Bakar, A.H.A. (2013). Analysis of the performance of domestic lighting lamps. *Energy Policy, 52*, 482–500.

7. Stan, A. I., Swierczynski, M., Stroe, D. I., Teodorescu, R., Andreasen, S. J., & Moth, K. (2014, September). A comparative study of lithium ion to lead acid batteries for use in UPS applications. In *2014 IEEE 36th international telecommunications energy conference (INTELEC)* (pp. 1–8). IEEE.

8. Erisman, J. W., Sutton, M. A., Galloway, J., Klimont, Z., & Winiwarter, W. (2008). How a century of ammonia synthesis changed the world. *Nature Geoscience, 1*, 636–639.

9. Fowler, D., Coyle, M., Skiba, U., Sutton, M. A., Cape, J. N., Reis, S., . . . Vitousek, P. (2013). The global nitrogen cycle in the twenty-first century. *Philosophical Transactions of the Royal Society B: Biological Sciences, 368*, 20130164.

10. Erisman et al. (2008), How a century of ammonia synthesis changed the world.

11. Iribarne, J. V., & Godson, W. L. (Eds.). (2012). *Atmospheric thermodynamics* (Vol. 6). Springer Science & Business Media; Yau, M. K., & Rogers, R. R. (1996). *A short course in cloud physics.* Elsevier; Wang, P. K. (2013). *Physics and dynamics of clouds and precipitation.* Cambridge University Press; Allen, J. T. (2018). Climate change and severe thunderstorms. In *Oxford research encyclopedias: Climate science.* Oxford University Press. https://doi.org/10.1093/acrefore/9780190228620.013.62.

12. Macfarlane, R. (2019). *Underland: A deep time journey*. W. W. Norton & Company.

13. Macfarlane (2019), *Underland.*

14. Van Valen, L. (1973). A new evolutionary law. *Evolutionary Theory, 1*, 1–30.

15. Strotz, L. C., Simoes, M., Girard, M. G., Breitkreuz, L., Kimmig, J., & Lieberman, B. S. (2018). Getting somewhere with the Red Queen: Chasing a biologically modern definition of the hypothesis. *Biology Letters, 14,* 20170734.

16. Harris, P. L., Koenig, M. A., Corriveau, K. H., & Jaswal, V. K. (2018). Cognitive foundations of learning from testimony. *Annual Review of Psychology, 69,* 251–273.

17. Sperry, D. E., Sperry, L. L., & Miller, P. J. (2019). Re-examining the verbal environments of children from different socioeconomic backgrounds. *Child Development, 90,* 1303–1318; Gilkerson, J., Richards, J. A., Warren, S. F., Montgomery, J. K., Greenwood, C. R., Kimbrough Oller, D., . . . Paul, T. D. (2017). Mapping the early language environment using all-day recordings and automated analysis. *American Journal of Speech-Language Pathology, 26,* 248–265

18. Hurst, M. A., Polinsky, N., Haden, C. A., Levine, S. C., & Uttal, D. H. (2019). Leveraging research on informal learning to inform policy on promoting early STEM. *Social Policy Report, 32,* 1–33.

19. For example, Eberbach, C., & Crowley, K. (2017). From seeing to observing: How parents and children learn to see science in a botanical garden. *Journal of the Learning Sciences, 26,* 608–642. Marin, A., & Bang, M. (2018). "Look it, this is how you know": Family forest walks as a context for knowledge-building about the natural world. *Cognition and Instruction, 36*(2), 89–118.

20. Rowe, M. L., Leech, K. A., & Cabrera, N. (2017). Going beyond input quantity: Wh-questions matter for toddlers' language and cognitive development. *Cognitive Science, 41,* 162–179; Walker, C. M., & Nyhout, A. (2020). Asking "why?" and "what if?": The influence of questions on children's inferences. In L. Butler, S. Ronfard, & K. Coriveau (Eds.), *The questioning child: Insights from psychology and education* (pp. 252–280). Cambridge University Press.

21. Yu, Y., Landrum A. R., Bonawitz, E., & Shafto, P. (2018). Questioning supports effective transmission of knowledge and increased exploratory learning in pre-kindergarten children. *Developmental Science, 21,* e12696.

22. Dou, R., Hazari, Z., Dabney, K., Sonnert, G., & Sadler, P. (2019). Early informal STEM experiences and STEM identity: The importance of talking science. *Science Education, 103,* 623–637.

23. Coffman, J. L., Grammer, J. K., Hudson, K. N., Thomas, T. E., Villwock, D., & Ornstein, P. A. (2019). Relating children's early elementary classroom experiences to later skilled remembering and study skills. *Journal of Cognition and Development, 20*(2), 203–221.

24. Callanan, M. A., Castañeda, C. L., Luce, M. R., & Martin, J. L. (2017). Family science talk in museums: Predicting children's engagement from variations in talk and activity. *Child Development, 88,* 1492–1504; Danovitch, J. H., & Mills, C. M. (2018).

Understanding when and how explanation promotes exploration. In M. M. Saylor & P. A. Ganea (Eds.), *Active learning from infancy to childhood* (pp. 95–112). Springer; Callanan, M., Legare, C. H., Sobel, D. M., Jaeger, G. J., Letourneau, S., McHugh, S. R., . . . Watson, J. (2020). Exploration, explanation, and parent-child interaction in museum settings. *Monographs of the Society for Research in Child Development, 85*, 1–137.

25. Schwartz, D., Bransford, J., & Sears, D. (2005). Efficiency and innovation in transfer. In J. Mestre (Ed.), *Transfer of learning from a modern multidisciplinary perspective* (pp. 1–51). Information Age.

26. Waters, T.E.A., Camia, C., Facompré, C. R., & Fivush, R. (2019). A meta-analytic examination of maternal reminiscing style: Elaboration, gender, and children's cognitive development. *Psychological Bulletin, 145*, 1082–1102.

27. Wu, Y., & Jobson, L. (2019). Maternal reminiscing and child autobiographical memory elaboration: A meta-analytic review. *Developmental Psychology, 55*, 2505–2521.

28. Shah, P. E., Weeks, H. M., Richards, B., & Kaciroti, N. (2018). Early childhood curiosity and kindergarten reading and math academic achievement. *Pediatric Research, 84*, 380–386.

29. For example, Callanan et al. (2017), Family science talk in museums; Dou et al. (2019), Early informal STEM experiences and STEM identity.

30. Weisberg, D. S., Hirsh-Pasek, K., Golinkoff, R. M., Kittredge, A. K., & Klahr, D. (2016). Guided play: Principles and practices. *Current Directions in Psychological Science, 25*, 177–182.

31. Crowley, K., Callanan, M. A., Tenenbaum, H. R., & Allen, E. (2001). Parents explain more often to boys than to girls during shared scientific thinking. *Psychological Science, 12*, 258–261.

32. Ridge, K. E., Weisberg, D. S., Ilgaz, H., Hirsh-Pasek, K. A., & Golinkoff, R. M. (2015). Supermarket speak: Increasing talk among low-socioeconomic status families. *Mind, Brain, and Education, 9*, 127–135.

33. Gawande, A. (2009). *The checklist manifesto*. Metropolitan Books.

34. Walker & Nyhout (2020), Asking "why?" and "what if?"

35. Agnew, D. C. (2010). Earth tides. In G. Schubert (Ed.), *Treatise on geophysics: Vol. 3. Geodesy* (pp. 163–195). Elsevier; Beeler, N. M., & Lockner, D. A. (2003). Why earthquakes correlate weakly with the solid earth tides: Effects of periodic stress on the rate and probability of earthquake occurrence. *Journal of Geophysical Research: Solid Earth, 108*(B8).

36. Hassinger-Das, B., Palti, I., Golinkoff, R. M., & Hirsh-Pasek, K. (2020). Urban Thinkscape: Infusing public spaces with STEM conversation and interaction opportunities. *Journal of Cognition and Development, 21*, 125–147; Bustamante, A. S., Hassinger-Das, B., Hirsh-Pasek, K., & Golinkoff, R. M. (2019). Learning landscapes: Where the science of learning meets architectural design. *Child Development Perspectives, 13*, 34–40.

37. Hatch, M. (2014). *The maker movement manifesto*. McGraw-Hill.

38. Halverson, E. R., & Sheridan, K. (2014). The maker movement in education. *Harvard Educational Review, 84*, 495–504.

39. Schad, M., & Jones, W. M. (2019). The maker movement and education: A systematic review of the literature. *Journal of Research on Technology in Education, 52*, 65–78.

40. Moutinho. S. (2021, March 3). Watch the winners of this year's 'Dance Your PhD' contest. *Science*. https://www.sciencemag.org/news/2021/03/watch-winners-year-s-dance-your-phd-contest.

41. Moore, M. (2015, Oct 29). WINNER "Dance Your PhD" Physics 2015: EnTANGOed—PhD researcher Merritt Moore. https://www.youtube.com/watch?v=BzKdKJn9El4.

42. See Cham, J., & Whiteson, D. (2017). *We have no idea: A guide to the unknown universe*. Penguin.

Index